CHANGING TRENDS IN ANTARCTIC RESEARCH

ENVIRONMENT & ASSESSMENT

VOLUME 3

The titles published in this series are listed at the end of this volume.

Changing Trends in Antarctic Research

edited by

Aant Elzinga

University of Göteborg,
Department of Theory of Science and Research,
Göteborg, Sweden

From an International Symposium in Göteborg, 30 September – October 1, 1991 to mark
the 30th anniversary year of the Antarctic Treaty, with a special concluding chapter based
on a recent evaluation of the activities of the Scientific Committee on Antarctic Research
(SCAR)

Springer-Science+Business Media, B.V.

A C.I.P. Catalogue record for this book is available from the Library of Congress.

ISBN 978-94-017-3761-6   ISBN 978-0-585-28849-9 (eBook)
DOI 10.1007/978-0-585-28849-9

Cover page picture: The picture on the cover is from the 1991/92 Swedish Antarctic Expedition to Dronning Maud Land in Antarctica, which was led by Olle Melander. It was taken during a traverse of the Kiberg Valley in Heimefrontfjella up towards the high plateau of Amundsenisen. It symbolizes modern compact technology which has become essential in polar research.

Printed on acid-free paper

© 1993 Springer Science+Business Media Dordrecht
Originally published by Kluwer Academic Publishers in 1993.
Softcover reprint of the hardcover 1st edition 1993

PREFACE

The core of this volume is a report from a symposium held at the University of Göteborg in the Fall of 1991. It deals with the interplay of science and politics and how such interplay affects research agendas. The focus is on polar research in Antarctica, a continent that has been much in the news during the past couple of years.

It gives me particular pleasure to thank all the speakers who took part in the program. All of them have many commitments and involvements in international polar research and the protection of Antarctica for its scientific and aesthetic values. The fact that such a distinguished group has been willing to come to Göteborg, to my mind attests to the importance and timeliness of our topic and the relevance of epistemological and policy issues in this field. A presentation of each speaker and author is made within the relevant chapters in the text.

My interest in the Antarctic has its origins in discussions with Anders Karlqvist, the Director of the Swedish Polar Research Secretariate at the Royal Academy of Science in Stockholm. Anders and I had worked together in the early 80's in a program on Technology and Culture, among other at the Research Policy Institute in Lund. At the time he was with the Swedish Council for Planning and Coordination of Research (FRN), its Committee for Future Oriented Research headed by Torsten Hägerstrand. Here we had many discussions regarding scientific disciplines as forms of culture and ways of life. This was about the same time that sociologists and ethnologists of science were bringing out a number of books on the tribal life and practices of scientists in laboratories. In these works, we found, the policy dimension was often treated in rather anemic fashion. The strong point of this newer literature was its focus on research practices, something which tended to be ignored by traditional philosophers of science. In Göteborg the program on Technology and Culture helped crystallize the present approach with its emphasis on the politics and epistemology of science.

The Polar Research Secretariate in Stockholm was created as part of a national effort that was to place a Swedish station in Antarctica. In moving to this new agency Anders brought with him an interest in historical and policy dimensions of science. This has been significant in bringing complementary perspectives - political science, history and theory of science as well as science policy studies - to bear on polar research. The library collection of books and documents collected at the Secretariate has been a useful resource for researchers in these areas, and I hope it will continue to expand.

I am grateful to Marika Lönnroth Carlsson, Information Officer at the Secretariate, for her help in various ways. She has supplied the picture which appears on the cover of the book. It is from the 1991/92 SWEDARP expedition to Dronning Maud Land led by Olle Melander, and was taken during a traverse of the Kiberg Valley in Heimefrontfjella up towards the high plateau of Amundsisen. It symbolizes compact modern technology which has become so essential in polar research.

Further, I want to record my appreciation of the many stimulating interchanges I have had over the years with Anders Karlqvist and others at the Secretariate in the course of my own pursuit of the polar. The Secretariate has also for my part, been an important vehicle for contacts with researchers in various natural science disciplines engaged in polar work. In the course of my work this has landed me in various conferences, from Luleå, to Oslo (Nansen Institute), Fairbanks (Polar Regions in Global Change), Brussels , Bremen (SCAR), to Geneva, as well as cocktail parties on the icebreaker Oden II upon her return form the North Pole, and Gotthilf Hempel's and the Alfred-Wegener Institute's unforgettable reception in their warehouse on the quay of Bremerhaven. I have also benefited from interviews with scientists in a number of countries, among other in Brazil and Japan, where for ten days I enjoyed the hospitality of the Japanese Polar Research Institute and wish here to thank its staff.

The symposium in Göteborg comes out of a project "Antarctica as a natural resource and object of research", funding for which has been made available by FRN under its commitment to the mandate of the Brundtland Commission. Uno Svedin, who has been a moving force at the Council in this connection has put together a volume together with Britt Hägerhäll Aniansson, entitled *Society and the Environment: A Swedish Perspective* (Kluwer, 1992). This was published just prior to the UN conference on the Environment in Rio June 1992, and used as a Swedish input to the discussion of sustainable development on our globe. Some of the results of my "Antarctica project" are presented there in a chapter. Further results will be found a chapter in the Sociology of the Sciences Year-book (Kluwer) for 1992 edited by Elisabeth Crawford, Terry Shinn and Sverker Sörlin. The topic there is internationalization and globalisation of science. The topic of Antarctic research as a form of Big Science is further developed in a contribution to a colloquium organized by the Royal Dutch Academy of Sciences (KNAW) in September 1989 (cf. E.K. Hicks and W. van Rossum eds., *Policy Development and Big Science*, North Holland Publ. 1991).

Within the same FRN-project I started a study of research profiles relating to different nations' use of science to manifest their presence in Antarctica. Results of this study will appear, in Swedish, in the report series of the Department of Theory of Science, the University of Göteborg. It will be the fruit of a joint collaboration between Lisbeth Johnsson (Department of Political Science) and myself. Lisbeth also exercised her organizational talents in helping make the symposium happen. On both scores I wish to record my thanks.

I am grateful to Ingemar Bohlin at my own Department for constructive views and help with the design of the symposium. He and I have collaborated on the politics and epistemology of science in an earlier project, which among other things led to the joint paper "The Politics of Science in Polar Regions" (*Ambio* vol. XVIII, No. 1, 1989 - pp. 71-76). Now we were able to draw on the network of contacts with Scandinavian scientists that he developed in the course of his review of Swedish polar research from the point of view of the theory of science and research (Cf. Ingemar Bohlin, *Om polarforskning*, rapport nr. 167 in series 1, Inst. för vetenskapsteori, Göteborgs universitet, 10 sept 1991).

The *Ambio*-article reappears, revised and updated, as the first chapter in the present volume, providing some background information which will help the reader more easily to grasp the context of the discussions at the symposium. The editors of *Ambio* are thanked for allowing me to reprint the article in its present form. The issue from which the article has been lifted is a thematic number on polar research that should be read in its entirety.

The very last chapter of the book is written by Rita R. Colwell, who chaired the international five-person panel appointed by the International Council of Scientific Unions (ICSU) to review the Scientific Committee on Antarctic Research (SCAR). This review document was presented to ICSU at its meeting in Jerusalem in early November 1992. The chapter appearing here is based upon the views outlined in that document. The chapter has been included here for two reasons. The first is that, as one of the members of the international review panel, I have been involved in the development of the original report. Secondly the review appropriately complements the discussion of the symposium in several respects. I am glad for the opportunity to publish these views here and thank Professor Colwell for her willingness to write a special chapter for the present book, with due acknowledgements to the other members of the panel for their input to the original report and ICSU for permission to use the substance of that report as a basis for the chapter as it appears here. Dr. Colwell, it will be noted, is currently Vice-Chairman of the Polar Research Board in the United States.

Finally I want to thank Anders Alvers at the Department of Theory of Science for his unflagging efforts and ever cheerful handling of the programs needed to produce a camera ready copy of this book for publication.

Four papers are included because they were made available to us in publishable form, for which I am grateful to the authors. In other cases promemoria have been used to fill out details in the summary of the proceedings appearing in each case under the name of the introductory speaker in the main body of the text. Jarl-Ove

KÄNNER DU INTE IGEN OSS?

"Don't you recognize us?", ask J.G. Andersson and S. Duse as they meet Otto Nordenskjöld 12th Oct 1903. Together with Grundén the two had spent a winter isolated in a stone hut where they burnt penguin oil for heat and light. Now they were all covered with soot as the leader of the 1901-03 Swedish Antarctic Expedition told them of the fate of the vessel "Antarctic" which had gone down under the ice seven months before. Sketch by S. Duse

Strömberg's notes in this context deserve mention because they were helpful in guiding us through the jungle of European acronyms on the polar research front; his text has been incorporated into the summary rather than printing it in part VI, thereby avoiding duplication.

It has not been possible to put together a complete proceedings of paper presentations. However, owing to the importance of the topics dealt with, and continual requests, we have decided to publish this abbreviated version in the form it appears here.

To introduce the reader to the interlocutors, there is a list of symposium participants included in appendix I; and a more detailed presentation of the authors is, as already noted, given successively in connection with each chapter of the book.

I have used the license of my function as rapporteur to select and structure the discussion in its present form. Sub-headings have been used to highlight various points and issues. Inaccuracies that may have crept in during this process are not to be attributed to the participants who have spoken, but fall upon my lot. So also does on the other hand the enjoyment of having pored over my notes (no tape recorder was used) to reconstruct the discussion, thus frequently discovering new dimensions and questions in what was said during the two days in the Fall of 1991. I hope some of this voyage of discovery will be conveyed in the final text as it appears here.

Göteborg in January 1993   Aant Elzinga

CONTENTS

PART IV
ORIENTATIONAL SHIFTS IN ANTARCTIC RESEARCH AGENDAS

PART V
PANEL DISCUSSION AND PLENARY

PART VI
FOUR SYMPOSIUM PAPERS AND A REVIEW OF SCAR

APPENDICES

Glossary

ATCP	Antarctic Treaty Consultative Party
ATP	Antarctic Treaty Parties
AWI	Alfred Wegener Institute for Polar and Marine Research
BAS	British Antarctic Survey
BIOMASS	Biological Investigations of Marine Antarctic Systems and Stocks
BIOTAS	Biological Investigations of Terrestial Antarctic Systems
CCAMLR	Convention on the Conservation of Antarctic Marine Living Resources
CCAS	Convention on the Conservation of Antarctic Seals
CEE	Comprehensive Environmental Evolution
CEC	Commission of the European Community
CEP	Committee on Environmental Protection
COMNAP	Council of Managers of National Antarctic Programs
CRAMRA	Convention on the Regulation of Antarctic Mineral Resoursce Activities
EC	European Community
ECOPS	European Committee for Ocean and Polar Science
EEZ	Exclusive Economic Zone
ESF	European Science Foundation
GOSEAC	Group of Specialists on Environmental Affairs and Conservation
IASC	International Arctic Science Committee
ICSU	International Council of Scientific Unions
IFREPOL	Institute Français pour la Recherché et la Technologie Polaire
IGBP	International Geosphere-Biosphere Programme
IHB	International Hydrographic Bureau
IOC	Intergovernmental Oceanographic Commission
IUCN	International Union for Conservation of Nature and Natural Resources
IWC	International Whaling Commission
JGOFS	Joint Global Ocean Flux Study/SCOR & IGBP
PEP	Protocol on Environmental Protection
SCALOP	Standing Committee on Antarctic Logistics and Operations
SCAR	Scientific Committee on Antarctic Research
SCOPE	Scientific Committee on the Problems of the Environment
SCOR	Scientific Committee on Oceanic Research
SO-GLOBEC	Southern Ocean-Global Ocean Ecosystems Dynamics Research
OS-JGOFS	Southern Ocean-Joint Global Ocean Flux Study
SOZ	Netherlands Marine Research Foundation
SPA	Specially Protected Areas
SSSI	Sites of Special Scientific Interest
SWEDARP	Swedish Antarctic Research Programme
TAAF	Territoire des Terres Australes et Antarctiques Françaises
UNEP	United Nation's Environmental Programme
WMO	World Meteorological Organization
WOCE	World Ocean Circulation Experiment / SCOR & IOC

INTRODUCTION

Sustainability implies development without incringing on the needs of the next generation. It involves some form of intergenerational equity, or at least an idea of this. At bottom then is a dimension of ethics, which in the case at hand translates into environmental ethics.

Sustainable Development

In the case of Antarctica, sustainability has to do with global ecological security. The situation of the continent and many of its physical characteristics both provide insight into the changes in the world climate in the past and a basis for understanding Antarctic's influence as a climate machine in the future. In addition it constitutes a strategic site for monitoring manmade disturbances of the environmental health of our planet. The depletion of the stratospheric ozone layer is only the most dramatic example of such a disturbance which has been detected there.

Important also for coming generations is that there will still be fish, that krill, which is the important resource at the bottom of a complex ecological chain, will not be seriously depleted. Further there is concern that the ice sheet may not be stable, that the greenhouse effect may cause it to melt and drastically raise world sea levels, with catastrophic consequences.

The conservationist guide-lines that are in place by virtue of the Antarctic Treaty put a ban on nuclear weapons and weapons testing in the region. They also prohibit the disposal of nuclear wastes. There is a special convention for managing marine resources, and recently the Environmental Protocol has been adopted which provides a comprehensive approach to the protection of a unique environment and will within the framework of the ATS, promote the institution of mechanisms to serve

1

this purpose.

All these developments have a significant bearing on science. Increasingly scientists have been called upon to serve in advisory capacity to governments in connection with resource management and environmental issues as these have come up on the political agenda of the Antarctic Treaty Consultative Party Meetings, ATCPM, the forum where ultimate decisions are made. Also science itself has been identified as a contributor to localized environmental impacts. In the future such impacts will be subject to prior assessment procedures.

Changing Perspectives

Both of these trends, the involvement of scientists in a political and bureaucratic decision-making machinery, and the setting up of environmental controls on research activities themselves, have implications for the social and cognitive conditions of science in Antarctica. These have until now not been the subject of serious study in themselves. Rather the debate on the pros and cons of science policy advisory mechanisms and the introduction of environmental checks on research stations has been coloured more by speculation than a reasoned review of experience or comparison with similar situations in other fields of science.

It is clear that changing perspectives on Antarctica as a natural resource will affect the perception of this icy continent as an object of research. This has been so in the past, and it will continue to be so in the future. The object of the present project with which this symposium is affiliated is to assemble some knowledge with regard to this complex interplay of the two dimensions - natural resource and object of research. We wish to explore changing agendas on the global policy arena on the one hand and the changing trends in Antarctic research on the other, cognizant of the fact that the latter are a product of both internal and external determinants, i.e., factors internal to the scientific endeavour itself as well factors impinging upon or setting the broader frame of reference and institutional context of the activities carried out at any given point in time. Thirty years ago when the Antarctic Treaty was established basic research stood supreme. Ten years ago political pressures associated with perceived prospects of oil and mineral potential, together with the actual needs of marine resource management, came to have a bearing on the activities of scientists, including the way research agendas were defined and defended vis a vis the taxpaying public. Presently it is the pressure from environmental organizations and the anticipation of a comprehensive conservationist regime where science comes under the scrutiny of impact assessments that has become a major concern. Some value it as a mixed blessing where good intentions might put an end to good science.

From Philosophy of Science to Science Policy

In the philosophy of science we have been concerned with these kinds of questions for some time. They form part of the problematic of science policy studies, an area where epistemology and politics can meet.

One question we have sought to bring to the attention of politicians and other decision-makers, as well as pressure groups outside science, is that there are certain minimal conditions and characteristics which must be in place for an activity to be able to continue as the thing we call science. External relevance and accountability pressures have in some cases tended to warp scientific enterprises to a degree where they become something else - in other words, research efforts as we see them require the maintenance of science-driven agendas and internal criteria of knowledge assessment (i.e. epistemic criteria). Under certain conditions such criteria may be overshadowed or crowded out by external epistemic criteria that have to do with the social or political relevance of the research in question or with the logistical operations of carrying it out. I have called this shift of epistemic criteria under external pressures from internalist to externalist focus, epistemic drift. It is a drift in the dynamics of the research process itself. Of interest for policy-makers, as well as scientists and citizens, is the closer study of what characteristics and circumstances are required to maintain the stability of the research process and therewith quality science in a multipartisan world.

To some extent this is a conceptual problem. Politicians and policy-makers, as well as external pressure groups, tend to have a simplistic image of science. In the 1960's science policy was dominated by the thought figure of a "science push" model. It was a time of expanding economies, and therefore the notion that if one primed the pump of science lots of good things would come out of it for society seemed to be a reasonable one. With economic decline, and the increase of social welfare programs, a "demand pull" or "market pull" (depending on one's ideology) model of the innovation process took over during the 1970's. Today there are attempts to combine the two approaches in what I call an "orchestration model", involving considerable degrees of freedom for individual scientists, but constrained by an overall framework that is steered on the basis of one or another social, economic or political mandate. Still there is a lack of sensitivity on the part of decision-makers when it comes to the complex interplay of social and cognitive factors that must be in place for research to do what it is supposed to do, provide us with a better understanding of how the world works.

To some extent it is also a practical problem of what institutional arrangements and organisational forms are most suitable for various kinds of scientific activities to be performed with maximum efficacy, internally as well as from the point of view of those outside science.

In our studies of the changing conditions of polar research we have found it timely to bring together a select group here to a round table to explore both conceptual and practical questions. To see where different interests have a common ground and to delimit more clearly areas of disagreement when it comes to the changing conditions under which polar research in Antarctica takes place and may be done in future. Hopefully this will contribute to a better co-understanding of the needs of both science and society.

The core of the symposium discussions appears in chapters two to six. These are preceded by chapter one, which is meant to provide some essential background information. At the end of the book I have also included a special chapter based on the views of the international panel that was asked by the International Council of Scientific Unions (ICSU) to review the place and function of the Scientific Committee on Antarctic Research (SCAR). This is a timely document that casts further light on some of the subjects touched upon in the present volume. It recommends that the structure of SCAR be examined and that a reorganization be considered. It also introduces the idea of an Antarctic Science Foundation, in itself international, intended to coordinate, stimulate and raise the funding needed for research in the Antarctic. This new institutional arrangement would fall under ICSU, and in affiliation with SCAR.

The book is thus structured so that the different chapters are grouped into several thematic sections or Parts, with the first giving an historical background to the Antarctic Treaty and introducing current concerns. This is followed up in part II by a review of the role of science, past and present. In part III changing trends in Antarctic research are then looked at in more detail, particularly considering the Environmental Protocol. A scientist's and an environmentalist's view are pitched against each other. Part IV puts the focus on ongoing shifts in Antarctic research agendas, while part V presents a concluding discussion of the symposium, and part VI assembles the four papers offered in publishable form. Finally, in this section a special chapter has been added, written by Professor Rita R. Colwell on the basis of the results of the international review panel that looked into the work of SCAR.

Sub-headings have been used in the text in order to highlight various points and issues. I hope this will help bring out facts, viewpoints and proposals in sharper relief, since the volume is intended to stimulate further discussion on Antarctica as a natural resource and object of research.

PART I
Historical and Contemporary Issues

CHAPTER 1

The Politics of Science in Polar Regions

by Aant Elzinga and Ingemar Bohlin

This chapter is meant to provide a backdrop for some of the issues taken up in the chapters that follow. Also, it presents some analytical terms that are useful in science policy analysis, particularly dealing with motivations and the practice of polar research, with special reference to the Antarctic. It introduces a concept of institutional motives, reviews some of the driv-

Aant Elzinga holds a B.A. in theoretical physics and applied mathematics (1960 - with a gold medal in mathematics) from the University of Werstern Ontario (Canada), an M.Sc. in history and philosophy of science (1964) from University College London (UK), and higher degrees in the theory of science and research (vetenskapsteori) from the University of Göteborg. He is presently Full Professor, holding a Chair in this discipline at Göteborg.

Ingemar Bohlin holds a B.Sc. (1985) from the University of Göteborg, and is a research student enrolled in the Ph.D. program at the Department of Theory of Science and Research. He has publishred several articles relating to topics in the history, theory and social studies of science, as well as a couple of reports on the structure and policy of Swedish polar research. His thesis work is in the historiography of science with special reference to scholarship on Darwin.

ing factors in modern polar research and considers some similarities and differences between Arctic and Antarctic science, in order to highlight the latter. External political conditions that form the framework within which polar research is done today, differ considerably in the two regions. In the Arctic the exertion of national sovereignty, as well as military and economic interests in a number of countries have hindered the far-reaching international cooperation in science found in the Antarctic. At the same time these factors have contributed to a fragmentation of knowledge production, while in the Antarctic, an international treaty arrangement which suspends territorial claims and emphasizes research has created conditions favorable to basic research. The focus is mainly on the tradeoff between science and politics in the Antarctic, and it is suggested that research there has a symbolic instrumental function, as distinct from a practical instrumental function which is most prominent in the Arctic.

The recent adoption of an Environmental Protocol (Oct. 4, 1991) for the Antarctic, now in the process of ratification, puts a ban of fifty years or more on all activities connected with mining and minerals exploration. By contrast, research relating to environmental concerns is strongly encouraged. This reorientation of institutional motives has had a significant impact on research agendas, in some cases implying a shift towards environmental monitering and practical instrumental modes of research. Therewith a major policy dimension has been reinforced in Antarctic research. Owing to the practical instrumental function involved, much can be learned here from general experience in the Arctic regions, where strong relevance pressures operate in many different spheres of endeavour, oftentimes to the detriment of basic research.

Institutional Motives

The environment has come to the fore as an issue that cannot be treated on a par with other issues. A recent Canadian report states: "The evidence suggests that human life itself is threatened unless we adopt a value system that places environmental integrity above all but the most basic human needs. This means treating the environment as a context for political and economic decision making rather than merely one consideration among many" [1]. When this fact is recognized the significance of polar research has to be underlined. The Brundtland report, for example, notes how the combined pressures of economic, technological, and environmental trends give urgency to the question of an equitable management of at least one of the poles. The Antarctic presents "challenges that may reshape the political content of the continent within the next decade" [2]. Polar research today stimulates the growth of science in many disciplines. In some cases it is motivated by human curiosity, or what may well be called the basic research motive. At the level of an institution or as part of a nation's science policy, the basic motive manifests itself in the existence and support to a group of specialists who define their problems at some distance from the daily pressures of politics and economics, frequently as part of a broader community bridging across national and international boundaries. This is only possible when the group is sufficiently large to maintain itself as a relatively stable and continuously active community, relatively immune from commercial, military, political and other socalled external pressures [3]. In this sense, the basic research motive is "internalist", as distinct from "externalist" institutional motives which have their point of departure in the quest for economic gain, national prestige, or power.

In polar research internal and external motives are more easily distinguished than in many other fields. One of the points of this chapter is to show why this is so, and how the various institutional motives function as driving forces, influencing the conditions of polar science differently in the Arctic and the Antarctic. Altogether we identify six motives in polar research. Their relative intensity will vary in different

Table one Institutional Motives in Polar Research [26]

1. *Basic Research Motive:*
   Many problems in basic research are studied with advantage in polar regions

2. *Political Motive:*
   Research activities mark national presence; successful polar research programs give international visibility and prestige for one's own nation. Science is used as a means to underscore and strengthen sovereignty claims or geopolitical stature

3. *Economic Motive:*
   A. *Natural resources:* in the Arctic there is a growth of resource extraction and production; the Antarctic constitutes a natural resource base, but with the Environmental Protocol having been adopted this has been closed as far as minerals is concerned. Marine resource exploitation is subject to regulation under CCAMLR, which is mandated to use an ecosystems approach
   B. *Technological development:* the development of equipment and science to exploit natural resources; new and emerging science and technologies are also interesting for scientists and military institutions

4. *Military motive:*
   A. From the *military* viewpoint the Arctic is a very useful battle arena in case of an armed conflict between the superpowers, because of the short distance by air and favorable conditions for missile-carrying submarines
   B. It has become important to defend *industrial* and other economically important installations in the Arctic region

5. *Jurisdictional/administrative motive:*
   The exercise of jurisdictinal functions and administration over societies, industries and natural reserves or conservation areas requires a knowledge base that must be continually developed

6. *Environmental motive:*
   The polar regions are suitable areas for probing and monitoring global environmental degradation (compare the ozone hole over the Antarctic), and global changes in the environment and climatic systems may be studied in geological time (e.g., through ice core drilling). The Antarctic is also relatively unspoiled and in need of protection

All six of these institutional motives prompt research in polar regions and help spur developments in a number of different disciplines and interdisciplinary specialties[27].

countries. In some cases the economic motive may dominate, in others the military, the administrative/jurisdictional, or perhaps the environmental motive (Table one gives a review of the different motives) [4]. It is possible that one might analyze and compare the polar research policies of different countries in terms of "motivational profiles", but this will not be attempted here. Our focus is polar research at the level of geopolitics. The emphasis will be on the Antarctic, while some reference is made to the Arctic to highlight a few comparisons. Being much more complex, the Arctic would require a much lengthier review than space permits.

The Changing Context

The politics of modern polar research reveals several distinct but overlapping periods of development. To some extent these are marked by changes in the driving forces. In the case of the Antarctic an early period of political tensions culminated in the mid-1950s. The cold war and conflicting interests amongst countries with territorial claims threatened to turn the continent into the scene of a free-for-all struggle. The International Geophysical Year 1957-1958, and afterwards the introduction of the Antarctic Treaty (AT) were instrumental in reducing tensions. Therewith began a new period marked by relative stability and a strong focus on science. The area was deemed a nonmilitary zone and the performance of substantial research qualified new nations for fullfledged membership in the "club" that is responsible for managing Antarctic affairs. Science acquired a symbolic value as political capital, and it flourished without too many external pressures of a more direct kind. During the 1970s however new issues emerged, prompted by economic and environmental motives. We are now into a third period in which the spotlight shines intensely on the Antarctic. During the 1980s there were many reports on minerals negotiations, expeditions, and preparations for new expeditions; this was followed by the environmentalist turn and the signing of a new Protocol. The situation is quite different from the one in which the Antarctic Treaty was drafted. Science is now required to contribute more systematically to a variety of external goals. Research to generate knowledge about the ozone hole, possible changes in the global climate, and for marine resource management are examples of the orientation.

In the Arctic tensions have continuously increased. From the late 1960s onwards they were aggravated by the scramble for off-shore oil exploration and development. Military motives and the complication of sovereignty claims associated with the extension of economic zones have also been sources of conflict, creating difficulties for scientific cooperation [5], basic data collection, and itineraries of research vessels. It is only very recently, with Glasnost in the Soviet Union and then the Russian takeover, that far-reaching East-West cooperation in Arctic science became a realistic possibility [6]. In practical terms it has meant the establishment of the IASC (International Arctic Scientific Committee), a partial Arctic counterpart to SCAR (the Scientific Committee on Antarctic Research), the latter is the international professional body that is responsible for research coordination in the Antarctic. (Table two indicates further features in a contrast of the contexts of Arctic and Antarctic science). A major new hazard in the Arctic is the nuclear waste material and discarded submarine reactors that have been dumped off the coast of Novaya Zemlya and elsewhere. Here the principles of national sovereignty and the concept of extended economic zones (EEZ) raise a hinder to a correct determination of the very extent of the problem today.

In the Antarctic the breakup of the cold war has taken away an important spur, as super power rivalry used to get translated into scientific competition. However the new political situation, together with budgetary constraints in both the US and Russia, has led to an increased interest in cooperation. This is evident both in the case of a joint project 1992 involving scientists from the two countries working together on a drifting research station on a large iceberg floating in the Weddell Sea, and in regard to a discussion of setting up a new research base in the interior of the continent.

Various factors have combined to strengthen the driving forces behind polar research. Industrialization of the Arctic and commercial interest in Antarctic minerals and hydrocarbons belong to the economic motive. Environmental consciousness has continuously grown since the 1960s. Thirdly, the movement of the nonaligned and Third World countries gathered momentum during the 1970s under the banner of a new international economic order. This has been responsible for putting the Antarctic on the UN agenda. Finally, there are the technological advances that have impacted on both science and the economy.

The Oil Crisis and Quest for new Resources

The economic motive gained momentum in the 1970s. The Club of Rome report with its pessimistic forecast relating to the depletion of global resources sounded an alarm in 1972. Rising prices because of Middle East political issues precipitated an oil crisis that sent energy prices rocketing, and the hunt for new resources was on. In this context the concept of Exclusive Economic Zones (EEZs) was evolved, and attention also turned to the polar regions. In Canada and Alaska the hunt for arctic petroleum, already intensified in the 1960s after scientific exploration, now held promise of payoffs. By 1980 over 100 coastal states (of a possible 139) had claimed some form of EEZs, pushing national economic boundaries far into the oceans. Though the Antarctic is still exempted from EEZ claims, several countries began making seismic surveys. Traces of hydrocarbons in the Ross Sea set off speculation about billions of tons of oil reserves. Major oil companies encouraged seismic surveys of the profile of the seabed, and expeditions have been carried out under various national flags to collect seabed data on the Ross, Weddell, and Bellingshausen Seas, and off the Antarctic Peninsula. The continental margin in the Weddell Sea area is regarded as promising.

Small finds of copper, uranium and platinum on the continent itself, and larger finds of iron ore and coal have further fuelled speculation about Antarctica as the

11

Table two: A Contrast Between Arctic and Antarctic With Respect to

Arctic

• *Geography:* Ocean surrounded by land; boundaries with non-polar regions not distinct

• *Biology:* Numerous animals and vegetation in the European as well as Asiatic and American parts

• *Indigenous population:* A number of tribes and ethnic variation in the European as well as Asiatic and American areas

• *Accessibility of natural resources:* Hard climate and sea covered by ice a major part of the year, but on the other hand there is a shallow continental shelf in many areas and the distance to industries and markets is relatively short

• *Current economic activity:* Exploration and exploitation of oil and gas on land and at sea is ongoing, as also mining, fishing and hydroelectrical power production

• *Military interest:* Extremely great and growing military interest

• *Sovereignty:* Sovereignty over land areas not subject to conflict, but boundaries between different countries' continental shelves and EEZ's unclear in various regions

• *International cooperation in research:* A certain amount of cooperation exists, mostly on a bilateral basis, but an International Arctic Scientific Committee created recently; research results subject to secrecy and only freely accessible to a limited extent

• *Relevance pressures on research:* In the *Arctic rim* countries research is strongly tied to military, economic and jurisdictional/administrative demands; in other countries the positions of basic research is strong

Features and Conditions that have a Bearing on Polar Research Activities

Antarctic

• *Geography:* Continent surrounded by oceans; distinct natural demarcation from other continents

• *Biology:* Animals ands vegetation only along the continental rim

• *Indigenous population:* No natural human settlements

• *Accessibility of natural resources:* Climate much harder than in Arctic; 99% of land surface covered by thick glacial ice; water off the continent very deep; distance to industrial centers and markets very long

• *Current economic activity:* Only fishing going on at present; moratorium on minerals exploration and exploitation for commecrcial purposes has been turned into a ban for at least fifty years, under the Environmental Protocol that is now in process of ratification

• *Military interest:* Demiliatarized zone, and today ony limited military interest; potentially of future interest

• *Sovereignty:* Seven counries have territorial claims, which are however suspended or resting by virtue of the Antarctic Treaty; the two superpowers reserve the right to raise claims in future

• *International cooperation in research:* SCAR successfully promotes cooperation in research at an international level, under the auspices of ICSU; demands under the Environmental Protocol (e.g., environmental impact assessments of planned projects) may facilitate further cooperation, as does the demise of former cold war tensions between the US and Russia. Scientists in France and some other countries have proposed the establishment of an international research base in the interior of the continent

• *Relevance pressures on research:* The position of basic research is relatively strong; strategic research has increased in volume during the past decade, and presently the environmental motive tends in some instances to press strongly in an applied direction on a spectrum of activities reaching over into environmental monitoring

world's last "treasure chest". The cost of drilling through the several-kilometer thick ice sheet to reach such deposits, the distance of transport to the nearest market, as well as the hazardousness of the enterprise make commercial ventures unlikely for many years to come. Still there was a tone of urgency in the minerals negotiations that took six years to complete, ending with the successful drafting of a document in Wellington in early June 1988. This new convention, open for signature from November 1988 onwards, did not come into force. Australia and France essentially put in their veto, and together with environmentalist lobbies across the world (in particular Greenpeace and the Antarctic and Southern Ocean Coalition – ASOC), they pressed for the introduction of a convention for comprehensive environmental protection of the Antarctic and associated ecosystems instead. In the political process that followed within the ATS this was turned into the Environmental Protocol, therewith turning a previous voluntary moratorium on minerals exploration (and exploitation) into a permanent fixture for at least fifty years. Within the Antarctic group there were divisions between prominers who wanted to exploit the continent, and those who wanted to put forth environmental protection as an overriding interest. Australia, New Zealand, Chile, and Sweden belonged to the latter category, while the pro-mining lobby included West Germany, Japan, the US, UK, France, and possibly Italy, and amongst developing countries Brazil and India [7].

Later France and Italy swung round to the environmentalist approach. The former USSR for its part implicitly belonged to the pro-miners. Now its interests have been taken over by the Commonwealth of Independent States, with Russia as the main actor. Also here environmentalist consciousness has developed. Moreover economic crisis has forced the Russians to cut down on their Antarctic effort.

At Wellington, Australia lobbied energetically at the minerals negotiations to get an antisubsidy clause written into the convention. This would prohibit artificial incentives and ensure that mining would not start until it is feasible by "normal" market criteria. This principle was blocked by the pro-miners, among them the US and UK. As critics have remarked, "it is somewhat ironic that several states which proselytize free-market economies were amongst those which most vehemently oppose the inclusion of the provision" [8]. Probably, the scale of future commercial ventures in the Antarctic is such that it is beyond the reach of individual companies, and would require joint ventures and backing from the public purse. Account was also taken of nonmarket economies like that of the former USSR.

The pro-miners scored a further point in obtaining a retreat from the requirement of full liability for accidents and environmental damage in cases where it can be "shown" that the source is unforeseen natural causes, armed conflict, or "terrorist sabotage." If and when this clause had ever been invoked in practice it would provide a lot of interpretative flexibility, and one could expect the power of mining companies and their recourse to hired experts to play an important role in the verdict.

It is this kind of expediency that reigned until a few years ago. Said one observer at the time: "... there is no evidence to suggest that the increasing concern of the Treaty members to develop a mineral regime can be attributed to long-term planning on their part. More probable is that various external pressures provided the principal impetus. The first of these is the fact that, by the early seventies, the world was seriously confronted with the importance of natural resources, their possible future shortages and their strategic values... This was followed by the sharp rise in oil prices . . ."⁹. There is no reason to believe that the pragmatism dictated by the promise of economic gain will diminish in the longterm future. In the interim however, Antarctic events have taken a dramatic turn whereby a ban has been placed on mining, as already mentioned, for at least fifty years. Australia and France have been instrumental in this connection, pushing for the line of turning the Antarctic into a "nature reserve – land of science". In connection with the new regime for environmental protection (Environmental Protocol), there will be a Committee for Environmental Protection in which scientists will play a central role.

In the Wake of the Law of the Sea

A second driving force that evolved during the 1970s is the one that culminated in UNCLOS III, the Law of the Sea Convention of 1982 (not yet ratified). The growing interest and rivalry in some cases over marine resources, seabed minerals and offshore energy showed up a sore spot. viz., the legal status of Antarctica. The Antarctic Treaty came into being with twelve signatories, including all the claimant states and both superpowers. Apart from science, with attention on the continent as a possible reserve for natural resources, several new nations saw an additional reason for joining the Treaty System, some of them placing research stations in the Antarctic in order to qualify for membership in the inner circle of decision-makers. Some Third World nations found the Treaty System completely unacceptable, viewing it as a remnant of big power politics and colonialism. Headed by Malaysia, a group of countries proposed an alternative concept for institutionalizing the management of the Antarctic, putting it under some kind of United Nations trusteeship. This is commonly referred to as the "common heritage of mankind" principle. Comparisons have been made with the common heritage monuments and sites that fall under Unesco. The idea may also be understood as an extension of the principles adopted in the UNCLOS convention governing deep seabed resources beyond the EEZs.

Once the UNCLOS text was settled the Antarctic was brought onto the UN agenda (1983), and several UN documents broached the issue of an alternative arrangement. Some countries asked for a postponement of the minerals negotiations until such a time that all nations might be consulted, and there was a request for UN participation in these negotiations, but this was ignored by the Antarctic group, even if the next meeting was deferred until after the UNGA debate. External pressures of this kind led to a certain accommodation both within the Treaty system and between it and outside interests. Acceding members for example had not been allowed to attend the consultative meetings where major decisions were made. In 1983, this rule was changed and acceding members began to participate with observer status. The

15

scientific criterion for consultative status was also weakened when India and Brazil upgraded their status in 1983, and in 1985 two further developing nations (China and Uruguay) became full members. These events helped defuse the Third World opposition. Nevertheless, the concept of a United Nations trusteeship on the basis of a "common heritage of mankind principle" continued to challenge the legitimacy of the existing regime and controversy has increased the visibility of the Antarctic issue.

Today, there are about 40 Antarctic Treaty members, 26 of them voting members. The majority of new members have joined during the past fifteen years, and today there are a number of different interest groups represented; industrialized nations, nations with territorial claims, developing nations, and the two superpowers who rank in a class by themselves. The strong pressure from some of the Third World countries has meant that in the Treaty System this now has to be taken into account. With regard to research these countries have disadvantages that may be compensated for by participation in bi- or multilateral joint ventures in science. Unfortunately, the tendency to retrenchment of national interests seems to mitigate against this possibility at a time when projects more and more have to be defined in global terms and linked up to international scientific initiatives.

Growing Environmental Consciousness

The environmental motive has a dual significance for polar research. It prompts linkage with global programs and integration in various disciplines. On the one hand research can contribute to protecting the polar environment; on the other the polar regions are good places to monitor the global environment and obtain data that are important for modelling physical, geophysical, climatic, and atmospheric systems. For global foresight, research from high altitudes and space into atmospheric aerosols and pollution over the two polar caps is of crucial importance. Measurement of trace elements in cores drilled in ice and snow are also significant. In the case of the Antarctic they afford a reference point (zero point) for the world's general environmental health sheet and its degradation in historical time. Variations in climate during the past 160 000 years can also be deduced[10]. Some of the data provide a basis for modelling variations in melting rates of ice covers and consequent changes in sea levels, which together with oceanographic studies and work in marine geology is important to the understanding of the Antarctic as a generator in a Southern Hemisphere climate system. At the same time research itself is also a cause of pollution. This is one reason why Greenpeace established an encampment (World Park Base - dismantled again in 1992), 30 km from the US McMurdo station and made several rounds of visits to a large number of research bases, during the austral summers, to determine the degree of pollution and environmental damage caused by their activities – toxic chemical wastes from photo laboratories, leakage of discarded oil drums, waste materials, dust and polystyrene particles, etc. [11].

Such findings have caused some sensation and led several countries to clean up some of the worst mess in order to improve their image. The non-governmental organizations have also been effective lobbyists-in-the corridors during the mineral negotiations, and they were instrumental with their watchdog activities to get the convention shelved in favour of an environmental regime. During the 1980s they have been an important source of information about the issues involved and events around such negotiations, and have been successful in finding media coverage for their own environmentalist and conservationist messages. By comparison, the advocates of the Treaty system have not been equally successful in public relations. In the eyes of many, the image of secrecy and exclusivity remains.

Environmentalists are bothered by the fact that Treaty members seem loathe to take up breaches of existing rules, as in the oft cited case of the French project to build an airstrip which runs right through a penguin rookery. This has never been taken up in official meetings – although of course unofficially pressure has been put on France to stop. Critics maintain that existing environmental rules lack the teeth needed for enforcement. Scientists point out that access to Antarctica by air greatly increased the length of the main research season and this is important for studies of the global environment.

Ecological systems in the polar regions are often said to be very fragile. This is however a point of some contention, as polar researchers in some cases point out how, owing to the extreme character of the climate, Antarctic ecosystems that have evolved are in fact fairly robust. Small changes in the environment will however have large effects, which is interesting for science. This does not detract from the message of the critics, who point out how footprints in polar lichen will leave their mark for years, while tractor tracks may leave irreparable gashes in the landscape.

Of course in the Arctic garbage from research stations is a minor problem compared to pollution around oil-drilling rigs and mining excavations, as well as industry in the Northern Hemisphere. Tourism is also a factor that has to be reckoned with – naturally there is a conflict between growing tourism and research. Scientists on King George Island, situated on the Antarctic Peninsula, and relatively easily reached by ships from South America, complain that frequent visits by tourists interrupt their work and also cause damage to the environment. The island is already overpopulated in terms of density of research stations. Regulation of tourism has become a priority issue [12]. The siting of research stations in accordance with scientific criteria rather than political expediency should be another.

The new Environmental Protocol stipulates that the environmental impact of all activities needs to be assessed prior to the inception of that activity. This includes tourism as well as scientific projects. A trilevel scale has been introduced to distinguish different degrees of impact, from marginal to major effect. Controls are to be stricter and more comprehensive the greater the anticipated impact of a given activity.

The Protocol has five appendices (four adopted at the Madrid meeting in October 1992, and a fifth added a little later at a meeting in Bonn). These may be changed on future occasions. The first appendix gives guidelines for environmental impact assessments. The second concerns protection of fauna and flora, while the third covers the regulation of waste disposal. The fourth appendix prohibits vessels from releasing oil, hazardous wastes and chemicals or the like into the southern oceans. The fifth appendix governs the regulation of "Specially Protected Areas" and "Antarctic Specially Managed Areas". The former include sites of special interest to science, or areas of specific aesthetic or historical value. Access to such areas requires special permission. Specially Managed Areas on the other hand are usually larger areas where human activities occur or may occur in future. A management plan is required for these, the purpose being to encourage planning and coordination, avoid conflicts, facilitate cooperation with other Consultative Parties and minimize risks to human beings and the natural environment.

The status of "Specially Managed Area" is decided at Consultative Meetings, and prior to this management plans have to be sent to SCAR, the Environmental Committee, as well as to the appropriate agency of CCAMLR. These three bodies can give advice, SCAR and CCAMLR for their part, it appears, only indirectly via the Environmental Committee. In the case of marine areas that may come under management plans, CCAMLR however has the final say.

The responsibility of implementing rules and regulations, and seeing to it that they are followed, lies with the national authority of the country under whose jurisdiction a project or activity is carried out, while the institution of an inspection system is to be international, but on a voluntary and multilateral basis. In practice one will probably see different national styles and standards in the implementation of different parts of the Protocol, as time goes on, and this may give rise to new tensions.

The Environmental Protocol, which constitutes an integrated and complementary part of the Antarctic Treaty, nevertheless takes its point of departure in the principle that the Antarctic must be regarded as a protected area, a natural reserve, to be utilized only for peaceful purposes, science and its unique aesthetic value as a wilderness. Mining and associated activities preparatory to mining in the future are forbidden. Previous regulations concerning the disposal of waste materials have been sharpened, and it is now the duty of each country to deliver a report on environmental and waste management at the research stations and expeditions that are run under its flag.

The adoption of the Protocol testifies to the fact that the Antarctic Treaty parties have now definitely changed their attitude regarding the previously proposed minerals conventions, rendering it obsolete. SCAR is recognized as an important player in upholding the environmental protection regime, although its exact role is somewhat unclear, which may tend to somewhat undermine its status. The Standing Committee on Antarctic Logistics and Operations (SCALOP), affiliated with SCAR, will also have new tasks under the Protocol.

Generally, it appears that the Environmental Protocol can provide a good basis for future cooperation between scientists and conservation groups. These could form new coalitions to pressure governments for more funding for commitments to environmental protection, so that these monitoring activities will not cut into the already lean scientific budgets.

18

Technological Advance

One factor that has done much to reshape the political context is high technology. Itself a product of scientific research and opportunity, technological advance has in turn been influential both in science and for the prospects of economic development and environmental ethics. The relationship is one of both cause and effect. In the north, polar scientific activities, combined with economic, military and global pressures, have been important driving forces for the development of many of the techniques which have helped change the nature of Arctic research [13]. Electronic sensors, satellites, computer modelling, techniques for deep-ocean drilling for geological purposes, new oceanographic equipment and polar research vessels have pushed science forward. New technologies have also opened up new possibilities in prospecting and resource exploration and exploitation of marine resources, minerals and hydrocarbons. Methods of ice- and weather forecasting have also been radically improved. Much of this technological revolutionization took place during the 1970s, spurring polar activities at a time when economic and environmental motives became stronger, and reinforcing these motives. In the Arctic military-strategic thinking has also been influenced [14].

This new era bears witness to many new concepts, some of them with a bearing also on international law and on legislation regarding resource management and impact analysis. Our vocabulary now includes such terms as artificial Arctic islands, seabed pipelines crossing "national" jurisdictions, offshore drilling platforms, freezer ships, under-sea oil, high-tech aquaculture, etc. They point to new forms of economic activities and new kinds of marine entities, as well as new rules of the game, in industry and international law. Because of their centrality in environmental monitoring, studies of weather and climatic systems, geophysical surveys and aspects of global change, many of these technologies, themselves emerging from science, have directly stimulated the various motives and helped set new agendas for polar research.

Globalization and the Internal Dynamics of Research

The increasing strength of the environmental motive in polar research, together with the advent of new technologies and developments in science more generally have accelerated a process of globalization of research. Individual projects if they are to be part of a research front are increasingly compelled to form part of global programs. An example is the International Geosphere-Biosphere Program under ICSU, commonly referred to as the "Global Change Programme." Here polar research plays an important role and ties into scientific work that transcends disciplinary boundaries. Rather than being a specialty in its own right, polar research more and more tends to become a recognized aspect in other disciplines. This is because the polar aspect is significant for understanding global systems. In some respects this is a realization of internationalist and global-scale concepts introduced in the Polar Years 1882-1983, 1932-1933, and the IGY.

What happens in the polar regions can affect the whole Earth in the long term. Freshwater runoff into arctic waters, for example, helps maintain a low salinity of the upper layer of water in the Arctic Ocean. This acts as a lid preventing warmer, deeper waters from reaching the surface, which would increase the heat flow and moisture transfer to the atmosphere. As a result global weather patterns would undergo drastic changes. Even a minor reduction in freshwater runoff may affect ocean currents and sea-ice formation and breakup [15]. Many scientists believe this will increase temperatures, but there are different views as to further effect. A "greenhouse effect" may raise sea levels, with disastrous effects, but it is also possible that increasing humidity will bring with it more ice which binds water, thus lowering seawater levels [16].

Industrial pollutants from Asia and Europe, and even from the southern US find their way into the Arctic, affecting vegetation, waters and the atmosphere. Projections of the "greenhouse effect" point to increased precipitation in the Arctic as an important factor to be studied. Changes in the ozone layer in polar regions may or may not be related to changes elsewhere around the Earth, but they are important for what they may show of planetary changes in energy balance, atmospheric stability, etc. The list of problems where the polar regions are important reference points is a lengthy one. The point is that we are seeing a globalization, both of environmental problems and concerns, and in the political arenas in which they must be dealt with. These two factors together influence the character of polar science, calling for international cooperation .

As already indicated, the globalization tendency in polar research also comes from within science itself. Polar research follows, and has in some important areas led the general trend in many other fields where systematic perspectives have been introduced – this is also a reason for national and international investments in Arctic and Antarctic science. In biology one now finds population ecology and behavior, ecological energetics, sociobiology and system-modelling. In molecular biology, research on the design of new life forms requires contextual approaches. In the earth sciences the theory of plate tectonics and sea-floorbed spreading provides a systemic and integrative basis for many previously unrelated fields, and some of the new techniques reinforce this trend. In the atmospheric sciences the discovery of the plasmapause 25 years ago, and the emergence of the field of solar terrestrial physics provides another example where previously unrelated studies can be linked to an overarching concept of the structure and dynamics of the interaction of two systems – earth (land, sea, ice) and space. The role of the Antarctic in the lithosphere, the formation of Gondwana and its breakup in geological time gives polar research a central place in, for example, marine geology. In climate creation models and global circulation models, and various other examples too, the poles are an integral source of data and concept formation in the broader disciplines: climatology, geology, geophysics, biology, and physical oceanography[17].

Historically, then, it might be said that polar research is now in a third phase of its internal dynamics, one signified by a globalization tendency and emphasis on theoretical work (and global modelling). The first phase was characterized by the taxonomic work of early days, which focussed on inventories of fauna, flora and natural features in the polar regions, a second phase (perhaps from the late 1950s) was characterized by a sharper focus on local processes. Of course, the modes of research associated with these earlier phases in polar research continue in tandem with the more modern approaches.

Applied and Strategic Research

The institutional response to the trend to globalization in polar research (both internally and externally driven) is varied. In the case of the Arctic, the strong linkages with economic, military, jurisdictional, administrative, and political motives has created a strong pull in the direction of applied research. Mission orientation tends to be tied to short-term problems and needs, leading to a fragmentation of knowledge production. This is a prominent feature in some of the Arctic rim countries (Canada, Norway, the US), while those countries that do not possess Arctic territories can concentrate their efforts on research of a more basic type (e.g. Germany and UK). Military and economic motives influence the thrust of research efforts. In many cases these are also tied to jurisdictional-administrative tasks, involving cartography, resource management, environmental impact studies, legal problems, and the extension of educational services, health care, communications, and cultural support for ethnic minorities, colonists, and migrant workers.

Concerning the Arctic, political analyst Oran Young has written, "it is no longer a frozen wasteland over which ballistic missiles would fly in wartime. The far north is rapidly industrializing and therefore becoming critically important to US and Soviet security"[14]. The mechanics of sea ice, polar clothing and housing, research on *aurora borealis,* and electromagnetic storms that interfere with defense systems or the operation of protective relay devices in oil pipelines – these are some of the many areas where military and economic motives converge. Other topics included under mission oriented activities are research on climate, natural hazards, modelling of glaciers, permafrost, energy-ice interaction, cold-climate engineering, cryosphere-ocean-atmosphere modelling, as well as problems in biological and physical oceanography.

Juan G. Roederer has pointed out how the strong links to application generates a fragmentation of Arctic science, ". . . many research needs in the Arctic are linked to political, economic and military interests. This makes international cooperation difficult; in certain geographic areas or fields of research, cooperation becomes impossible, when research findings are classified as sensitive or proprietary by governments or industrial firms. Even within a single country Arctic research issues may prove socially or politically sensitive. The research needed to resolve such domestic conflicts as impact of industrial development on Arctic environments, or the impact of western influences on native culture and subsistence lifestyles, cannot fail to stir strong political controversy in the Arctic communities most concerned"[18].

In the Antarctic the relevance pressures are not so strong. Mission orientation there takes the form of a long-term targeting of basic research[19]. We get what is frequently referred to as "strategic research"[20]. Roederer gives the following definition; "Strategic research is research with a possible long-term (say 25 years) payoff in applications. Most, if not all, of the input arises from research originally geared to improving knowledge generally."

It constitutes efforts to orient "the development in a given discipline toward achieving a predetermined but restricted goal in scientific knowledge"[18]. External research motives are translated and internalized into the agendas of basic research programs.

In the case of the Antarctic, then, the position of basic research is generally accorded a more central place than in Arctic science, and mission orientation when it occurs takes on the form of targeted or strategic research. This is because of the political context. Among the payoffs are increased knowledge regarding climate change, tectonics, resource management, and other applications pertinent to low latitudes. The introduction of the Environmental Protocol and the current emphasis on monitoring activities may however in some cases bring with it strong pressures that tend to undermine the status of basic research. This is a concern that has been voiced by various groups of scientists. Worrisome too is the thought that attention to environmental protection in future may be perceived to give better political payoff than a focus on pure science.

The Trade off Between Science and Politics

In the Antarctic, because of the treaty, which suspends territorial claims and makes science the ticket into the club of decision-makers, research represents a form of symbolic capital [21]. There is a special kind of tradeoff with politicians, whereby scientists are provided with funds to do research, but in doing this research they also perform a political task, advancing the national interests of their own country in a geopolitical arena. In doing so they can influence the growth of science. Crudely put, one might say that politicians don't need to worry so much about the kind of work their scientists do, as long as they are there in Antarctica and one can show that a "significant performance of research" is going on. The symbolic value lies primarily in the very presence of a country's scientists on this cold continent, but of course international recognition of highly qualified science enhances the symbolic value of a country's research on the political arena[22]. Probably, with time the latter aspect becomes more important, but then again this can vary from one nation to another, depending on the prevailing political climate, the national science policy doctrine and overriding motives. In some cases, a country may desire to join the club to influence the course of international science.

External relevance and accountability pressures easily distort scientific priorities. In the Arctic rim nations this is a major science-policy problem, but we find references to it also in the Antarctic; "politics plays a major part in Antarctic affairs and it must be expected that political considerations will inevitably intrude, in some countries, in the selection of which scientific activities to support. On purely scientific grounds, the evidence of the past 20 years suggests that in some cases a great deal of money has been spent in some programs supporting poor scientific objectives, or activities which are poorly coordinated and apparently represent no planned or systematic effort to fill in clear gaps in our knowledge"[23]. In some cases, then, the rhetorical import of research activities may be more important for politicians than their actual scientific value. Thus, the image of letting the scientists more or less follow their own heads (and hence the natural prominence of good quality basic research) does not always run true. Given the fact that polar research is very expensive, and that budgets are limited, while the urge to do something in the Antarctic may be great, there is a breeding ground for opportunism. The tendency to drift from internalist to externalist criteria in the evaluation of polar-research projects must constantly be guarded against in the scientific community. A conflation of criteria tends to undermine quality of results; since internal evaluations are downplayed as quick and immediate "use" is prioritized, theoretical work, important in the long run, suffers [24]. The stronger the externalist instrumentally driven motives, the clearer the polar scientist will have to be about the distinction between internal and external criteria for setting priorities of scientific fields. Around the externalist motive it is not uncommon to find subcultures of scientists mixed with planners, politicians, administrators, bureaucrats, and businessmen. These "hybrid" research communities form their own value patterns, criteria, modes of evaluating results, and possibly reputational systems and career patterns, which in part are distinct from those of the academic research communities centered on disciplines. The latter adhere to the ethos of the Republic of Science, which gives primacy to peer review and the determination of research agendas ideally in interaction with colleagues across the world in unconstrained dialogue. In fact, there are power struggles between different schools of thought; but there is still a common interest in basic research. In the "hybrid" communities, on the other hand, external determination of research agendas and the primacy of social or political relevance belongs to the order of the day. Intertwining of academic and "hybrid" communities is increasing.

Scientists belonging to disciplinary communities are not always happy about their roles as specialist advisers at meetings where bureaucrats and politicians dominate. The secrecy surrounding diplomatic negotiations also runs against the grain of the scientific ethos with its ideal of a free exchange of information. There is also irritation over the way the consultative parties of the AT have more or less come to regard SCAR as "their" scientific secretariat, putting a drain on already meagre financial resources[25].

In polar research the criteria and social control mechanisms are particularly important because of the extreme costs involved. Climatic conditions and logistics, payload costs, unusual modes of transportation, and extreme demands on equipment and maintenance, as well as on the reliability of measuring instruments, all of these call for exceptional care and rigor in the selection, planning and implementation of scientific programs. This means that internal rivalries between different schools and between academic and "hybrid" communities may become acute.

SCAR's Double Role

The division between basic or curiosity motivated and strategic research is reflected in SCAR's committee structure. On the one hand SCAR, as a member organization of ICSU, facilitates information exchange, communication and encourages cooperation in Antarctic research programs. On the other hand, on its own initiative or by request, it interacts with the AT system and provides important input to its meetings. The latter is an interesting example of the emergence of "hybrid" research communities. This is not an uncommon phenomenon in mission-oriented fields of science; in the case of Big Science it is particularly prominent.

Still SCAR facilitates a consolidation of basic research. The creation of a similar body for Arctic polar science may help offset the much stronger drift to applied and mission-oriented science that exists there.

Disciplinary groups in SCAR exist around specialties like upper atmosphere physics, biology, human biology and medicine, oceanography, etc. – nine groups in all. These follow the disciplinary boundaries in science, and their problem agendas are internally generated within the international scientific community. In addition to this SCAR organizes *ad hoc* groups of specialists. Their character is interdisciplinary and they form the loci of hybrid communities, their problem agendas being influenced by externalist motives. The topics suggest that their mandate includes a fair amount of strategic research: Antarctic Climate Research, Environmental Impact, Sea-Studies, Seals, and Southern Ocean Ecosystems. These are the names of the *ad hoc* groups, whose problem areas appear to be prompted by environmental motives, resource managerial objectives and knowledge needed for the implementation of conventions, among them one on the conservation of Antarctic seals (from 1972), and another called the Convention on the Conservation of Antarctic Marine Living Resources (CCAMLR of 1980). The latter has its own scientific advisory group. With the new Environmental Protocol, we have already noted a special advisory committee of scientific and technical experts will be set up, once the regime is established.

Exploitation and management of renewable resources (e.g., living stocks, and in future – icebergs) and previously, nonrenewable resources (oil, gas and hard rock minerals) has motivated strategic research in many different fields. This included foresight regarding the possible effects of large-scale exploration and exploitation, as well as more basic research concerning plate tectonics, mineralization processes, sedimentology and stratigraphy.

In connection with marine-resource management, SCAR earlier initiated a program for advancing knowledge of the structure and dynamic functioning of Antarctic marine systems. This strategic research program went under the acronym BIOMASS, or program for Biological Investigations of Marine Antarctic Systems and Stocks, which was organized in the mid 1970s and completed it's mission with a SCAR symposium in Bremerhaven in 1992. It coordinated work done in many different disciplines, but the ultimate objective has been the management of living resources which are and may in future be commercially exploited.

Concluding Remarks

When the Antarctic Treaty came into effect in 1961 the focus was on peaceful uses of the continent and international scientific cooperation. This is still the case today, although various pressures have increased, reflecting first economic and then environmental motives. This has led to the introduction of a couple of management concepts challenging the present AT-regime.

The Stockholm Declaration of the United Nations Conference on the Human Environment 1972 is sometimes taken as symbolic of the conservationist trend. Article No. 21 of this declaration stipulates that states should take measures to prevent damage of other states or of areas beyond the limits of national jurisdiction – of course, several governments deny that Antarctica is beyond national jurisdiction. In line with this principle emerged the idea of designating Antarctica as a "World Park". In 1975, New Zealand tabled this at a meeting of the consultative parties to the AT. Later some environmentalist organizations used the idea as a battle cry, to mobilize opposition to the Treaty system and work for an alternative arrangement different from both the Treaty and the "common heritage of mankind" concept. More recently some of the principles from Stockholm 1972 have been reiterated in Rio 1992, while in the Antarctic a minerals convention has been ditched in favour of an Environmental Protocol. This has lain the groundwork for new alliances, this time between scientists environmentalists.

As we have indicated, this and other events in recent years have increased the visibility of the Antarctic and research there. As yet, basic research still has a very strong position, thanks to the nature of the present political regime. This is significantly different from the situation in the Arctic, where the exercize of sovereignty, economic exploitation, military expansion, and various other factors pull science strongly in applied directions, with serious fragmentation in the production of knowledge as a result. Conscious science policies are needed in both polar regions to maintain a growth of science that will benefit a majority and not just a minority of mankind; the environment is one outstanding issue around which, hopefully, unity might begin to be achieved.

Notes

1. Science Council of Canada. 1988. *Water 2020. SCC Discussion Paper No. 40.* Ottawa, p. 23.
2. Brundtland Commission. 1987. *Our Common future.* The World Commission on Environment and Development. Oxford, 287 p.
3. Roederer, J.G. 1978. University research. Competition with private industry? *The Northern Engineer 9*, 26-31.
4. Further development of the idea of institutional motives is found in Bohlin, I. *Om polarforskning* (se ref. in Preface above p ix).
5. In a recent paper, Willy Østreng gives a very interesting description of the rationale for scientific cooperation in the polar regions, based on the concepts of symbolical and practical utility of science which I. Bohlin has developed in the articles cited in notes 4, 21 and 22. Østreng, W. *Polar Science and Politics Close Twins or Opposite Poles in International Cooperation.* Paper to the International Symposium "The Management of International Resources: Scientific Input and the Role of Scientific Cooperation," Oslo 10-12 October, 1988. A short version of this paper has appeared as "International Cooperation in the Polar Regions: the Role of Science", *International Challenges* (Oslo) *8, p.* 20-25.
6. A chain of events was sparked off by Gorbachev's October 1987 Murmansk speech (Pravda 2 Oct. 2. 1987) – a meeting of experts from Canada. Denmark/Greenland, Iceland, Norway, Finland, Sweden, the US and USSR met in Stockholm March 1988, and a scientific conference in Leningrad in December 1988 were steps on the road to the formation of an International Arctic Scientific Committee .
7. Sander, K. Greenpeace, Copenhagen. Report on the final session of the Antarctic minerals convention negotiations (personal communication).
8. Antarctic and Southern Ocean Coalition. 1988. *ECO (Wellington NZ) 63 p. 1.*
9. de Wit, M.J. 1985. *Minerals and Mining in Antarctica. Science and Technology. Economics and Politics.* Claredon, Oxford University Press, 53 p.
10. Genthon, C., Barnole, J.M., Raynard, F., Lorius, C., Jouzel, J., Barkov, N.l., Korotkevisch, Y.S. and Kotlyakov, V.M. 1987. "Vostok ice core: climatic response to CO_2 and orbital forcing changes over the last climate cycle", *Nature 329,* 414-418. This is the third of a series of three articles in *Nature*, presenting the results of French-Soviet research cooperation; the other two articles on the ice core studies are in *Nature 329,* 403-407 and 408-413, respectively.
11. Greenpeace. 1988. *Expedition Report 1987-88. Greenpeace Antarctic Expedition. Stichting Greenpeace Council.* Lewes, East Sussex, UK.
12. Rocha-Campos, A.C. (Secretary within SCAR). Geosciences Department, University of Sao Paulo, Brazil (personal communication – interview) .
13. Roots, E.F. 1986. Introduction. In *Advances* in *Underwater Technology. Ocean Science and Offshore Engineering.* Graham & Trotman publ. vol. 8. (Exclusive Economic Zones).
14. Young, O. 1986. The age of the Arctic. *Oceanus 29* 10-17, which deals with new technologies in the military sector.
15. Science Council of Canada . 1988. *Water 2020. SCC Discussion Paper No. 40* Ottawa, p. 16.
16. Olausson, E., Department of Marine Geology, University of Gothenburg, Sweden (personal communication – interview) .
17. Fifield, R. 1987. *International Research in the Antarctic.* SCAR/ICSU Press, Oxford.
18. Roederer, J.G. 1978. University research. Competition with private industry? *The Northern Engineer 9*, 26-31.

19. Walton, D.W.H. (ed.).1987. *Antarctic Science*. Cambridge University Press, 250 p.
20. For a discussion, see Irvine, J. and Martin, B. 1984. *Foresight in Science. Picking the Winners*. Francis Pinter, London.
21. Bohlin, I. *The Motive Structure in Contemporary Polar Science*. Paper presented at "The Study of Science and Technology in the 1990's", joint conference of the Society for Social Studies of Science and the European Association for the Study of Science and Technology, Amsterdam, November 16-19, 1988.
22. Bohlin, I. "Modern polarforskning. Anteckningar om dess samhälleliga roll", *VEST. Tidskrift for vetenskapsstudier* (Göteborg) 8, 25-35. (In Swedish)
23. Walton, D.W.H. (ed.). 1987. *Antarctic Science*. Cambridge University Press, p. 61-64.
24. Further discussion in Elzinga, A. 1985. "Research, bureaucracy, and the drift of epistemic criteria", in *The University Research System. The Public Policies of the Home of Scientists*. Wittrock, B. and Elzinga, A. (eds). Almqvist and Wiksell International, Stockholm, p. 191-220.
25. Walton, D.W.H. (ed.). 1987. *Antarctic Science*. Cambridge University Press, p. 59.
26. Box 1 presents a condensation of material in I. Bohlin cited in note 4.
27. For a survey of the various fields of science that are spurred by jurisdictional, economic, military and other motives, see *Arctic Research in the United States 1,* (Fall 1987) and 2 (Spring 1988).
28. The work behind this article has been made possible thanks to a grant from the Swedish Council for Planing and Coordination of Research (FRN). The authors also want to thank Fred Roots and an anonymous *Ambio* reviewer for many valuable comments on an earlier version of this article. We are also indebted to Peter Beck for discussion and comments on other work in this project. Ingemar Bohlin has given a more detailed presentation of motive structures in polar research in the report cited in note 4.

PART II
The Functional Role of Science in the Antarctic Treaty System

PART II

The Functional Role of Science in the Antarctic Treaty System

CHAPTER 2

The Role of Science in the Negotiations of the Antarctic Treaty–
an Historical Review in the light of Recent Events

Historical Background

Finn Sollie recounted how in late December 1958, the situation was one where the atmosphere of the cold war dominated the world. This also affected the "Antarctic question". Traditionally interest in the continent had been generated by its

Finn Sollie was intimately involved in the drafting of the Antarctic Treaty which appeared in 1959 and was adopted by a group of twelve nations in 1961. In his presentation he gave some personal reflections on the background behind this event and a glimpse of the dramatic late night negotiations in which scientists played an important role. His major point was that science in fact was the crucial element that made the treaty possible. Without science there wouldn't have been an Antarctic Treaty.

potential as a source of seals and whales that could be hunted, with furs and oil yielding a good price on a world market. Science had come in with these interests in resource exploitation. With the sharpening of the tensions founded in the surge of nationalism which triggered the First World War, the quest for making and defending territorial claims took over. Behind this were both economic and political motives: whaling and bases for whaling stations, as well as in the British case the right to tax whale oil. Of course an element of national prestige was also involved.

Even a country like Norway, which was not really interested in making territorial claims on the Antarctic mainland, was finally forced to do so in 1939 to preempt a nazi German claim to Queen Maud Land and the coastal territory surrounding it. These had been Norwegian whaling waters, and Norway wanted to assure itself of access. Originally Norwegian policy had always been to want open access for everyone - the principle of maintaining a commons. This was similar to the American position; the US for its part had adopted a principle of only recognizing effective occupation as grounds for sovereignty claims. Still in the late 1930's the US had developed a hidden agenda of encouraging its scientists, for example during the Byrd expeditions, to

Territorial Claims in Antarctica

Territorial Claims in Antarctica

20°W	-	45°E	Norway (upper and lower limits undefined)
45°E	-	136°E	Australia
136°E	-	142°E	France
142°E	-	160°E	Australia
160°E	-	150°W	New Zealand
150°W	-	90°W	Unclaimed
90°W	-	53°W	Chile
74°W	-	25°W	Argentine
80°W	-	20°W	United Kingdom

drop canisters with documents that might be used as a basis for future claims, just in case.

At the time the British government put forth a claim in Antarctica, in 1908 covering the area of the Peninsula, the Norwegians, newly independent, lacked knowledge of taxation and other legalities attached to questions of territorial rights in what was formerly a no mans land in the southern ocean. The supremacy of the British Empire and its experience in such matters was no doubt an advantage for the British. Norway with its policy of wanting to see a commons rather than claiming territory for itself thus did not get into a fundamental conflict with Britain over claims. Chile and Argentine however did about thirty years later.

From World War to IGY

With the Second World War a military strategic interest came into the picture. Traditionally interest in the Antarctic had been economic, scientific and political, as well as administrative and legal when territorial claims had been laid (legal and licensing knowledge was needed, which also stimulated research, e.g., in the Norwegian case). The Germans now began to ply the southern oceans with their submarines, and the British were afraid that they might set up a base there. Moreover when Britain was tied up in the war in the northern hemisphere, Chile and a German-oriented Argentine challenged the British rights to Antarctic territorial sovereignty by putting claims to territory in what the British regarded as their "sector". Early British activities of setting up stations in the region of the Peninsula were directly motivated by an effort to shore up the British claim, and likewise for the Chileans and Argentinians. By the end of the war the US also acted on the basis of a strong military strategic interest in the Antarctic.

The new power of the airforce made the continent more accessible. With the new atomic bomb technology the military was also interested in this far away region as a possible place to test the bomb. Further, strategists were reckoning that the next arena for a large scale war would be over the Arctic where North America faced the Soviet Union. This prompted military exercises to train in cold climatic conditions, not only in the Arctic, but the Antarctic was also interesting. Finally, with the advent of nuclear reactors the Antarctic became a place where nuclear waste material might conceivably be dumped.

These were the new interests that coloured discussions regarding a political regime for Antarctica. The US with its basic idea of keeping the continent open for Allied countries launched a plan for a Condominium of the seven claimant countries plus one (the US). The USSR was to be excluded. The plan backfired. It was opposed by the British, its Commonwealth allies, Australia and New Zealand, but also the USSR, which started to increase its presence with whalers in the southern oceans. During the early and mid-fifties the "Antarctic question" had become politically insoluble - a number of countries were at loggerheads with each other, over claims

(Britain, Chile, Argentine) and military strategic interests (especially the US and USSR). In this contest science turned out to be the common ground where nations could meet.

The International Geophysical Year 1957/58 was important not only for the science that came out of it, but also for the way it helped defuse political tensions over Antarctica during the cold war era. The stalemate from 1948 onward was broken.

Scientific Cooperation and Diplomacy

Finn Sollie pointed to the British-Norwegian-Swedish expedition to Maudheim as a model for international cooperation. It had some significance for the IGY cooperation. It was among others Lawrence Gould, "the father of Antarctic science", who pointed to the Maudheim model. The fact that the USSR prevented IGY cooperation in the Arctic made it even more important in the Antarctic.

Fortunately those who were involved in the Antarctic treaty negotiations were motivated by international interests more than short-sighted national ones. The Pentagon was very interested in keeping Antarctica open for nuclear tests, but this was averted during the last night of negotiations, not least thanks to opposition from Argentina, Australia and New Zealand. The US had a program for using nuclear power, civil as well as military tests, in Alaska, so the competence for moving into the Antarctic was there.

It is evident from the text of the AT how the nuclear and atomic bomb issue was a major one at the time. Science however became the pivotal point, and it was the requirements of science that set the conditions: free access to Antarctic territory, free use, free exchange of informations, allowance of inspection of any base by any member country's scientists, joint planning and execution of activities. All these scientifically oriented principles went against the grain of the legal and political demands of nations. Thus the Treaty defined a unique international regime.

Instrumental in the negotiations was Paul Daniels, who orchestrated eight or nine months of meetings of a nonpolitical group (scientists and administrators). This group found the road that led to the principles of 1) peaceful uses only; 2) freedom of scientific exchange should continue as under the IGY; 3) territorial claims should neither be affirmed nor denied, which was the same principle applied during the scientific activities of the IGY, that these activities should not be taken to affirm or deny claims; 4) a unique consultative process, with routine meetings at intervals to adopt recommendations that when approved by governments would become law. In this process scientists had a tremendous influence in those days, since diplomats had very little knowledge concerning Antarctica and its affairs.

An insightful presentation of the early history of the treaty is found in the book by Richard S. Lewis and Philip M. Smith, *Frozen Futures* (New York, Quadrangel, 1973) put out by the Bulletin of Atomic Scientists and the New York Times jointly.

Natural Resource Interests

The question of natural resources came up in Tokyo 1970 and Wellington 1972. In 1973 the Nansen Institute in Oslo hosted a private meeting on the question of minerals and mineral rights, which was an issue so difficult that you could not make official statements about it at the time. Marine resources came up in 1975; it was also difficult, but solved first (CCAMLR). CRAMRA took more time to hammer out. Here the Law of the Sea Conference was important, and so was the role of scientists. The former because it charted out new legal concepts and machinery to cover a mixture of claimed and unclaimed areas on the globe, and the latter because of the expert input they provided as a basis for decision-making.

Some Facts

Antarctica constitutes 10 % of the Earth's land surface
Only 5% of the coast is free of ice in summertime.
Distance to

South America ca.	1,000 km
South Africa ca.	3,700 km
Australia ca.	4,000 km.
Extension of continental landmass is	12,393,000 km²
Extension including mass of ice	13,975,000 km²
Ice free area	200,000 km²
Maximum thickness of ice cap	4,335 m
Average height of ice above land	1,720 m

If all ice were to melt the world sea
levels would rise by   over 60 m
Length of Transantarctic Mountain Range   4,100 km
Highest point on the continent
   Mount Vinson   5,140 m
Lowest measured temperature
Vostok research station (21 July 1983)   -89.2° C

Ave temp is on the average 30° C colder than in Arctic

Katabatic winds can reach up around 300 km/hr

CHAPTER 3

Development of the Science/Politics Interface in the Antarctic Treaty and the Role of Scientific Advice

The Political Dimension

Nigel Bonner noted how some expeditions early on, fired mainly by nationalist ambitions and prestige, tended not to yield much in the way of scientific results. The Amund-

Nigel Bonner has been involved in Antarctic science for a long time. Currently he is the Convener of the SCAR Group of Specialists on Environmental Affairs and Conservation (GOS-EAC), in which capacity he plays an active role in discussions pertaining to both environmental and science policies. In his presentation Bonner reviewed the history of Antarctic exploration and the advances made in science.

sen expedition was explicitly named in this connection. We were also reminded how British revenue accruing from taxation on whale oil production was used to finance research, especially hydrology and biological research pertaining to the welfare of whaling stocks in the southern ocean. The British *Discovery* expeditions which led to an accumulation of much data and new knowledge was basically motivated by resource management concerns. Thus the economic motive fed the scientific one.

During and after the Second World War politics became the main determinant. Britain was prompted to protect its military strategic interests and counter Chilean and Argentinian actions. In 1943 Operation Tabarin was launched, a military intelligence operation which was transformed into a civilian operation, the Falklands Islands Dependency Survey. After the war FIDS in its turn gave birth to the British Antarctic Survey (BAS), signifying the increasing prominence of science as lead activity as time went on. The process was not a smooth one; in 1952 bullets were fired by Argentinians over the heads of British scientists. The battle between the rival interests may also be traced in the place names of various landmarks and rocks, straits, and the like on British, Chilean and Argentinian maps, one country contesting the place names attached to such landmarks by the other countries and vice versa.

Geophysics came to be the discipline that drew all scientific subjects together in

Antarctica in the 1950's. This was with the IGY when 47 research stations were set up south of latitude 60°S, and another 8 north of this line. The formation of what became the Scientific Committee for Scientific Research (SCAR), under the auspices of the International Council of Scientific Unions (ICSU), gave scientists an international coordinating body. This was an important element in the evolution of the Antarctic Treaty that was drafted in 1959. The Treaty continued the practice of the IGY and turned it into a principle, viz., freezing the situation with respect to territorial claims and guaranteeing the freedom of scientific information.

Cross section of Antarctic icemass

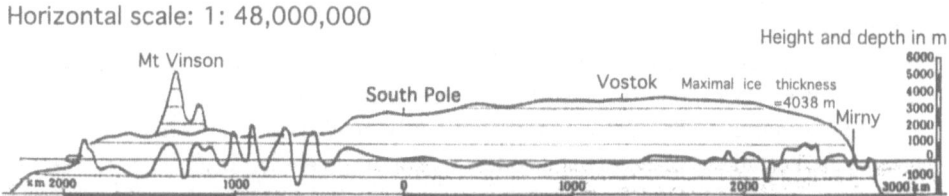

During the IGY a lot of effort went into mapping the profile of the continent underneath the icecap, which was much thicker than expected

Environmental Protection and Resource Interests

The Treaty in turn laid the ground for environmental protection. Measures for the protection of the environment are not explicitly stated in the Treaty, but implicitly there is an environmental conservationist philosophy, manifested for example in the prohibition on the dumping of nuclear waste materials. In 1964 in Brussels a set of measures were explicitly adopted to protect fauna and flora. These were based on a set of conservationist principles developed within SCAR. In the ensuing years concern with the protection of the environment increased, with SCAR playing an important advisory role. The scheduling of meetings with SCAR members getting together even numbered years and the Antarctic Treaty Consultative Parties during odd numbered years was an arrangement that facilitated constructive interaction. During these years it was often a small core group that took the lead - New Zealand, the United Kingdom, Norway and the US formed the core.

Interest in natural resources, both marine and mineral resources in the 1970' s led to an increase in the number of members party to the AT, a development particularly noteworthy in the 1980's. This latter was at the same time the decade during which environmental consciousness became a major factor in Antarctic affairs. Greenpeace turned its attention to the southern hemisphere and eventually placed a four-man station in Antarctica. The AT parties in the meantime struggled with the minerals question, which was anathema to the environmental groups affiliated with ASOC and Greenpeace. They succeeded in destroying the minerals convention, Nigel

Bonner maintained. Therewith a new series of conflicts developed, with discussion focused around what form environmental protection guide-lines should take: a convention (Australia, France, Belgium, Italy wanted this), a protocol (the US and UK position), or something in between (New Zealand). Undiplomatic maneuvering on the part of the US delayed the negotiations that led to an environmental protocol.

The Decline of SCAR

This development has led to a weakening of SCAR's role. Science has been given a less central role compared with what it had when the Treaty was created and during the first couple of decades. This decline of the position of SCAR coincided with the creation of a special advisory organ affiliated with CCAMLR, and now the new protocol on environmental protection will have its own advisory body, the Committee on Environmental Protection (CEP), which is basically political. Also the logistics group, the managers of Antarctic science, COMNAP, have been given a stronger position. According to Bonner, scientists have every reason to worry about the effects environmental regulation will have on science. "It is not paranoid for scientists to worry about the way regulation affects science", he said, and gave several examples where bureaucrats who have never been down in Antarctica have condoned regulations that are contradictory and hinder regular scientific activities. If they had at least some acquaintance with Antarctic conditions such stupidity could have been avoided.

Another problem is SCAR's double role. On the one hand it is to coordinate scientific research, on the other it still has the ambition and mandate to serve the Treaty organisation with scientifically based advice. Scientific delegates at Treaty meetings however have to speak through the heads of their delegations, which means that controversial issues may be toned down and factual knowledge in some sense subordinated to political issues as these are defined by politicians and legal experts. SCAR is in a double bind here. On the one hand the involvement in policy advisory capacity constitutes a drain on the resources and energies of scientists who might be better off to devote their time to basic research problems; on top of that, "advice" tends to be watered down to suit political agendas. On the other hand if SCAR were to retreat from its advisory function it might become even more marginalized, and lose influence, giving bureaucrats and politicians even greater leeway. Also there is the danger that the Environmental Committee which will be set up to assess impacts of scientific research activities on the environment might not even ask for advice from SCAR.

SCAR currently also lacks the infrastructure to respond effectively to all the requests for advice that come along. The annual budget is only about one quarter of a million dollars, which does not carry far when one has to bring together scientists from various locations in the world. The storage of data and policy regarding the handling of data is another outstanding issue that is under review, requiring better resources.

Perhaps the introduction of a permanent AT-secretariat might help alleviate some of the infrastructural problems, giving SCAR a chance to play a more leading role *vis a vis* other institutions within the ATS. There is some support for this idea. The provision for independent advice to the Treaty is most important. Failing such advice international diplomats may become even more the captives of populist ideas and media manipulations.

The general conclusion was that SCAR has unfortunately been losing influence over the years. The introduction of a new order where it is explicitly stated that Antarctica is a natural reserve intended for peace and science is important, but the regulations that are put in place have to be developed in close dialogue with working scientists, and SCAR is the body which speaks for science. Therefore SCAR has to be strengthened.

Nigel Bonner mentioned how in the original draft of the environmental protocol SCAR was given a stronger role than in the final product. There was a watering down of the wording regarding the role of SCAR. This reflects the general trend.

The discussion following upon this presentation took up the question if SCAR should retreat from its "political" role. Within the scientific community there is some opinion that goes in this direction. The source is in part the frustrating situation where scientists cannot give independent advice, since they have to speak through their government delegation's heads at the ATCP meetings. If one took the consequences of this the option of an alternative where SCAR only dealt with basic research may be an interesting one. Also there are some governments that have explicitly stated that they don't like non-governmental organizations to get involved in government matters. At ATCP meetings it is increasingly a question of countries gaining political advantage, which is not always in the best interest of getting good scientific advice. The best advice is not facilitated within present structures. The question is what other structures are conceivable.

Different Perspectives on the Future of Science

At the same time SCAR itself has in recent years discussed strategy for Antarctic science and raising its own visibility. In the course of this greater consciousness regarding policy matters ought to have developed within the community of Antarctic scientists, who are now, among other thanks to global programs, starting to integrate their research activities and knowledge production more closely with research fronts in some of the mainstream disciplines and specialties. It can be said that Antarctic science is being globalised, conceptually and in terms of research policy.

The ambition to raise SCAR's visibility outside the scientific community has not been very successful. This was evident at the Bremen conference "Antarctic Science - Global Concerns" 23-27 September 1991. This conference had been organized by SCAR with an explicit intent to reach out to environmentalists and the mass media, in order to dispel misconceptions about Antarctic activities and display the wide range of research work that is going on to contribute knowledge on some of the most pressing issues of time - global change in the Earth's climate and other important parameters. As it turned out the Bremen conference was not well attended by persons from outside the scientific community.

A question that came up in Bremen too was to what extent environment as a global concern can replace East-West rivalry reflecting at bottom military strategic interests, or the lure of minerals and hydrocarbons, as an external stimulus for the funding of research activities in Antarctica. Scientists are divided in their opinion regarding this question. Some maintain that such external stimuli never have been the prime motive force behind good Antarctic science. Others agree that economic, political and other external interests are important as a general framework and that the end of the cold war may imply less of an incentive on the part of governments to use science as a vehicle for upholding a presence in the region. The quest for minerals and hydrocarbons is generally not seen as an equally strong motive factor in the past.

Finally there are the optimists who believe that the new level of environmental consciousnes is a new effective force.

CHAPTER 4

Relevance Pressures and the Strategic Orientation of Research

External Determinants on Polar Research

In his presentation he made a distinction between internal and external determinants that propel science. With regard to the first the relationship between theoretical work and empirical observation is significant. In addition the relationship of Antarctic science to other scientific fields and specialties is an important determinant factor that contributes to shaping research agendas.

> Anders Karlqvist directs the Swedish Polar Secretariat which has a broad mandate relating to polar *research* including planning, coordination and implementation of Swedish polar research, diverse operations and management, as well as responsibility for the logistic function of Swedish polar expeditions. In this capacity, and with a background in applied mathematics and systems analysis, Karlqvist has for some time reflected on the interplay of different motive factors in Antarctic science and what bearing these may have on the performance of research.

With regard to the second category of determinants - external ones - two different contexts were identified: the political context and the operational context. In the latter various determinants are at work in the planning and execution of expeditions and the like. These appear in different forms, but one way to deal with them is to speak in terms of various criteria that must be observed in the operational mode. Five criteria were listed with regard to operations that define the thrust and content of expeditions:

- ethical acceptability of an operation
- safety of the persons carrying it out
- cost effectiveness
- technical feasibility of the operation
- environmental soundness (degree of impact an operation will have on the physical environment)

With regard to the political context, Karlqvist expressed the view that with the end of the cold war and environmental concerns replacing earlier political ones, external pressure to do research for its own sake (i.e., to be present), is partly fading. This also goes for the mineral resource interest which was significant in the previous decade; it also functioned as an external pressure to do basic research. With the environmental protocol this has been terminated.

The Environmental Turn

With the environment replacing the old military-strategic, political and environmental concerns, we now have a situation, Karlqvist predicted, where the political expression of Antarctic influence will take other forms. Science is no longer the pivotal factor it has been within the framework of the ATS. This is already evident in the way the placing of a research station is no longer the rule of thumb to get into the ATCM's or SCAR.

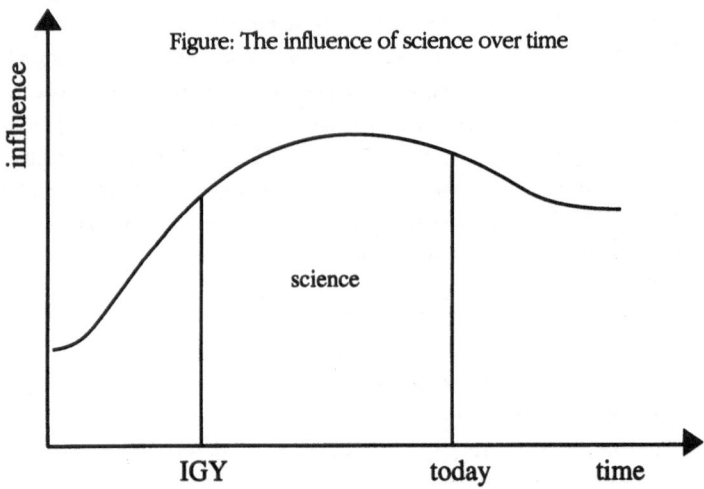

Figure: The influence of science over time

Today we are more concerned with obtaining results, getting answers from science when it comes to environmental questions that have a direct bearing on political negotiations between governments regarding pollution control and ecological security on this globe. Thus scientists are asked to provide accurate statements as to future sea levels, at what rate the Antarctic ice cap may melt, or if it is stable, what is the extent of the ozone hole and what does this mean for human and animal life, what traces are there of chemicals and other man-made pollutants in the polar atmosphere and cryosphere. This means that monitoring and mission oriented science will dominate over the traditional internally driven forms of basic research, even though there is , and will continue to be, room for basic research in Antarctica (e.g. solar research, astrophysics on the South pole, etc).

We will see fewer scientists operating in the Antarctic, relatively speaking and more journalists, authors, tourists, etc. The decline of the Antarctic field worker is in part also due to the advances made in high technology. In future electronic sensors and observation via automated stations, robots and satellites will replace *in situ* human observers. Thus, high tech and an expansion of tourism is the scenario Karlqvist saw, and he asked, what will this mean for politics and science. The speculation was that the influence of science would decline (see figure on the previous page); perhaps the fate of SCAR is a reflection of this general trend which is being accentuated with the environmental protocol.

Science Squeezed from two Sides

Coming back to the overall perspective, Karlqvist thus pictured science as being squeezed in from two sides, politics on the one hand and operations/logistical considerations on the other.

politics ⟶ science ⟵ operations/logistics

The differentiation in these dimensions is also institutionalized within the ATS, where there is a division of labour between the ATCM, SCAR and COMNAP.

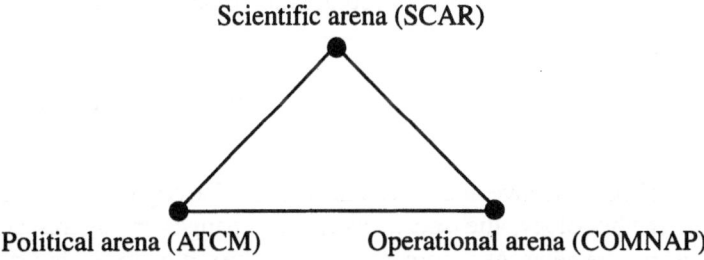

Scientific arena (SCAR)

Political arena (ATCM)   Operational arena (COMNAP)

We have already looked into the political arena, where the shift has been over to environmental concerns, with the named repercussions on the scientific arena. Consider now more closely the scientific arena. What is the relationship between theoretical work and empirical observation, and how does Antarctic science relate to disciplinary science on the one hand and socially mandated science on the other? Is it possible to discern any trends?

Once that question is answered the focus can go further, to the operational arena.

The Internal Dynamics of Polar Research

With regard to the internal or intrinsically driven determinants, Karlqvist found that Antarctic science has started later, is younger as a scientific field, and has basic empirical work to do before it can reach the maturity of the science of easily accessible areas. The tendency is that empirical work has dominated. It has driven theoretical work, and not the other way around. This is an inductivist mode of research. However, recently efforts have been made to reverse the process, so that data accumulation is more strongly tied to hypotheses and conceptual work at the level of theory. This is brought about in at least three ways. On the one hand it is a movement that is driven by the need to integrate consciously with international programs like the International Geosphere Biosphere Program (Global Change) where data from Antarctica play an important role. On the other hand, at the subjective level, SCAR's efforts to develop a strategy for coordination and planning of research guided by theoretical work at the research front in various areas of mainstream science is significant. Thirdly, the technology factor, with sophisticated computer modelling of ocean circulation, atmosphere and cryosphere, and the attempts to link these to General Circulation Models and climate models evokes reorientations in a direction of model- or theory-steered research work. Altogether it may be said that the "modernization" trend in Antarctic science thus leads to a situation where research agendas will be more theory-driven. Thus the trend is from empirically driven to theory-led research.

This process may be schematically represented as follows:

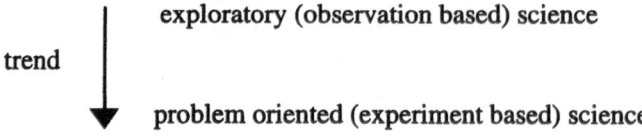

This trend has both cognitive and social dimensions. The cognitive have to do with changes in methodology, the social ones with changes in organisational and institutional arrangements. At a more general level the same process is seen to have double components. On the one hand there is a closer integration of Antarctic science with disciplinary science, so that it becomes less clear what the "Antarctic" aspect is; the concept "polar research" also becomes fuzzier.

Tensions in Antarctic Research

On the other hand there is a pressure to integrate Antarctic science with socially mandated non-Antarctic science. In recent years this trend has been overtaking the pressure to integrate with disciplinary science. This is owing to the overall changes on the political arena where environmentally related efforts and other activities, like tourism, are becoming more central (compare the earlier remark regarding a shift from science to monitoring). A problem is that Research Councils have a mandate to fund science, and these new activities are not always seen as scientific, while at the same time funding for environmental impact assessment and the like is not immediately forthcoming from other agencies. This places an extra burden on scientific budgets.

The increased concern with the environment is also reflected in criteria that must be taken into account in the operational context. In addition to questions of technical feasibility of projects and cost effectiveness, during the past decades there has been more emphasis on the welfare and safety of personnel, and now in addition to this environmental impact assessment has emerged as an explicit factor with its own criteria.

Karlqvist maintained that the criterion of ethical acceptability of projects is a marginal consideration when compared to other areas of modern research, such as medicine; of course ethics comes into the picture when animals are caught or the behaviour and health of groups of Antarctic researchers are studied.

Cost Effectiveness and High tech

With regard to safety, there has been a long term trend of new restrictions and constraints on research. Most countries do not accept the adventurous escapades and dangers associated with expeditions during the heroic age. Indeed with present day standards Scott and Shackelton, for example, would not have been allowed to embark on their famous exploits. Since the crash of the New Zealand DC-10 on Mt Erebus civilian passenger flights over Antarctica have been stopped, and there is a much greater consciousness concerning safety precautions. Also with the age of high tech and sensors connected to satellites the romantic element of the earlier dog sled era is definitely gone. Of course a certain tension between low tech and high tech approaches continues to exist. However robotics and automation permits us to collect data in places where no human being can tread (e.g., remotely guided submersibles under the ice shelf).

Cost effectiveness is a criterion that has always been important in Antarctic research, since logistics takes up such a large proportion of budgets. Today we see a mixture of Big Science, middle sized science and even some small scale field operations. The general trend has been toward what John Ziman has called "bounded science", i.e., a situation where a ceiling prevents continual expansion of R&D allocations. In the case of Antarctic science resource constraints are hitting operations very hard at the same time as the development of sophisticated equipment has made operations more capital intensive.

Technical feasibility is a criterion that pressures us in the direction of smaller, faster and more high tech oriented operations. At the same time the newer technology, with

45

automated data collection, electronic instrumentation and reliance on satellites tends to favour the current reorientation toward more mission directed research and monitoring activities. This also permits more focused and less opportunistically defined projects.

Environmental Impact Assessments

The introduction of environmental impact assessments and the idea of multi-use management plans means that environmental criteria will be given a much more prominent place in polar operations. Many of the cases of pollution and degradation of the Antarctic landscape around research stations are the accumulation of old sins, and these have to be dealt with, besides determining the proper conduct of ongoing activities. In some cases it is more a question of aesthetic blemish than environmental degradation in the real sense of the term. The newer stations are being built on quite different premises, with waste recycling, and now of course a lot of garbage has to be shipped back to its countries of origin. This is an additional strain on research budgets.

The question of environmental criteria takes on a different form, depending on which level one poses it:

– political (public opinion, other pressures)
– bureaucratic implementations of (new regulations)
– practical work (dealing with a backlog of old sins and guiding
  the proper conduct of ongoing activities)

In the end it is the scientists and expedition personnel in the field that determine if or not operations are executed in an environmentally sound way; the concept "environmentally sound" also leaves some leeway for interpretative flexibility.

Monitoring and Safety

The discussion following Anders Karlqvist's presentation came to focus on two main issues, monitoring and safety.

Why are scientists so reluctant to talk about monitoring? Why has it become a dirty word in some scientific circles?

One reason seems to be that Research Councils do not support activities that cannot be identified as research. Monitoring falls outside the scope of their regular funding mandates.

Another reason is internal peer pressure and career patterns in science. Monitoring may not yield the kind of results that look good in a scientific paper to be used to advance one's career. Academic work, including the PhD dissertation, requires data collection to be placed in a context where problems are generated from within an internal scientific agenda. There is a difference here between career lines that stay within the mainstream of academic science, and those that link up with regulatory agencies and the like. A distinction might be made between basic or disciplinary scientists and hybrid research scientists, where the latter have a double filiation, both academic and non-academic. If monitoring can be tied to a theoretical problem in science it will get more credibility; the combination of monitoring and globalisation within the context of international programs like the IGBP may perhaps contribute to a partial transformation of value systems.

Several participants insisted that ethical acceptability was a criterion that does come into play in Antarctic science. For example, who do you sell your services to, and your results? This is not an insignificant question. Bioethics also comes into the picture - in connection with the treatment of penguins and other animals that may be subject to monitoring. Safety is also an ethical consideration. Americans have prohibited climbing in mountains or going out on sea ice under certain conditions; the British are more relaxed, because their philosophy is that such restrictions would put a curb on the enthusiasm of younger men who still seek some adventure in Antarctic science.

Early mode of transport. Sketch by S. Duse

47

General Discussion

The general discussion revolved around a review of different external determinants that have driven and still drive science. It also came back to the question of monitoring.

It was pointed out that resource exploitation has not ceased to be an external driving force for science. In the case of marine resource exploitation the need to manage resources is generating a lot of studies. Here different approaches to setting limits to the catch of krill can have a bearing on the role of scientists as policy advisers and hence indirectly on scientific agendas.

Another driving force for a number of nations was 1991. This was the magic year when many believed the Antarctic Treaty might be up for review, and hence one had to get onto the band wagon before it was too late. This prompted some nations to launch Antarctic expeditions and set up a research station so as to qualify for membership in the ATCP group which makes the decisions concerning the continent's future. Some third world countries saw a potential mineral resource treasury that they did not want to be excluded from. Now the "1991 question" is gone, and a protocol that puts a ban on mining is in place, so these driving forces are largely gone for the time being.

The Changing Conception of Science

The orientation toward environmental concerns has brought with it a "drift" in the conception of science itself, Anders Karlqvist maintained. In many cases one finds Antarctic science comprising a component of some large scale international research program, for example dealing with the southern ocean. In this case it is not a question of monitoring programs, but process studies of a different sort than traditional ones. Their results may lay the ground for monitoring, but should not be confused with monitoring as such. Thus we have here a conception of science that is halfway between traditional basic research and monitoring.

The Dutch example of joining the ATCP group without mounting a research station but relying on excellence in marine science confirms the shift in the definition of the "science criterion" in another way. This does not mean that the demand for "substantial research" has changed, only that the form of this research, because of the new context, allows for new options.

One of the participants drew the conclusion from this discussion that it is important that scientists promote the notion of what "science" really is, and what it is not. One has to persuade bureaucrats of the relevance of Antarctic science, as distinct from monitoring. The latter may be geared to empirical data collection efforts, but the more long term challenge posed by global change calls for advanced theoretical work and modelling. This must be made clear for bureaucrats and politicians. The argument of science for the good of your soul or national prestige does not bite, if it ever did.

Funding Structures

Nigel Bonner noted how external funding from research councils and sectoral agencies, in as far as these require frequent project reviews, may tend to drive science away from long term programs. Efforts tend to concentrate on smaller projects which may not have the continuity that one wants in order to build up a long term fund of knowledge or develop deeper theoretical work. It can be bothersome to have to be concerned with peer review every time you turn around; in some respects a concentration of resources in a national polar research institute is better because it avoids this problem and allows for longer term continuity in larger scale programs.

This is an interesting point in the Swedish context where there is a tension between an academic career system based on discipline research and the need to pull together research efforts in terms of a "polar" or "Antarctic" effort over a longer period of time. The Japanese Polar Research Institute, which is an example of the opposite model indicates that a large concentration of resources in an institute may reduce flexibility. When new problems come up on polar research agendas it may be more difficult to reorient one's national efforts because there is at the same time a certain inertia in the core of competence one has built up over a long time.

New Technologies

Several participants agreed with Karlqvist that new technology was an important driving force in Antarctic science. Jan Stel saw a "technology push" at work (remote sensing, for example), and maintained that scientists in the 1990's will need new tools. At the same time there is a difficulty here, because some of the new tools generate more data than computers and the human beings can handle scientifically. This technology push consequently increases the need for focused research and theoretical or model-driven work.

The tendency of Antarctic science going high tech was not unanimously accepted as either good or necessary. Several participants objected to the picture of an automated Antarctic science with researchers sitting at their computer desks in their home countries reading printouts of data via satellite transmissions. Barry Heywood insisted that modelling is only a tool for science. It does not reduce the need for field scientists. On the contrary these are crucially important in order to determine baseline data. Moreover the discovery of the ozone hole should be a lesson. It shows the need of field research, creative individuals who can come up with new angles on things. The human individual is the irreplaceable factor in Antarctic science. Another example is the fact that ships are needed down in Antarctica to see what happens to the krill; a satellite can't tell you that.

Another point that came up in the same vein was the need of scientists to change the programs for formatting data in other areas, for example in the study of the movements of krill in order to determine stocks. Automatic data collection generally only takes the parameters that have been programmed for; this is a sore point behind some of the controversies surrounding general circulation models in climatology, because there are disagreements concerning what parameters to ignore or how to scale down or up when moving between specific and more general models.

Those who were more enthusiastic about the possibility of automation in Antarctic research made a comparison with the situation in space, where robots do a lot of the work in space observation platforms. This analogy was immediately criticized and rejected by others who pointed out how first of all the space effort in many cases was much less science and more a question of technology development or commercial ventures. Secondly, the point remains that there are a lot of things one can't do via instruments, and also simple technical faults can create havoc unless one has a scientist "down there" (or "up there" in the case of fixing the space telescope, for example) who can change the program of research so that one can still get a lot of scientific value out of a modified program. Why would one otherwise send technicians down to Antarctica, if it were not for the fact that a lot of unforeseen things happen, instruments have to be fixed, trouble shooting has to be performed on the spot. Failing this, Antarctic science would be much less cost effective.

Also, of course there are differences between different fields of science. What goes for marine science, for example, may not work for geology. In geology one still has to go out into mountain ranges with a simple hammer. Moreover scientists must always determine the relevance of data. As for automation, it should be steered by hypotheses and modelling work - this was a view around which there was consensus.

Anders Karlqvist ended this part of the discussion by clarifying his position. He had only meant to emphasize one point, viz., that we are faced with a future where priority setting will be more technology-driven than ever before. This is not the same as glorifying the role of high tech.

Monitoring - Scientific and Otherwise

Olav Orheim opened the discussion on monitoring once more. It was noted how the word monitoring can mean quite different things to different people. Perhaps a distinction should be made between routine observation for long series of measurements over time, data collection for hypothesis testing and finally monitoring which is driven by externally determined tasks. Thus some people use the term "monitoring" for checking if something is wrong, for example if the number of lead particles in the atmosphere have increased. This is not science.

Ozone measurements are a form of monitoring. In itself such an activity has a bad reputation among scientists, it is not considered good science. Funding bodies are also unwilling to commit themselves to long term programs of this kind. However if such activities are connected to a scientific problem the situation changes. Thus one and the same activity may be "science" in one context and "monitoring" in another.

Several participants confirmed the perception that monitoring is not something one pushes forward when applying for money from research councils, and not in peer reviewed periodical literature either. Still, even in science, observations are needed over long periods of time to establish changes in various parameters with a degree of certainty. Instruments are tools that need to be developed - there is a need for progress in instrumentation both for science and for extrascientific purposes.

Large scale collection of data series also requires a cooperation between scientists from many different countries, and the development of instrumentation calls for closer cooperation between scientists and engineers. Thus the technology factor, seen in this perspective, is also a driving force with respect to globalisation and internationalization of research in Antarctica.

James Barnes was critical of the tendency to underestimate the value of monitoring that he felt some scientists expressed. In his view monitoring had been given a bad "rap". If it is the case that monitoring has a bad reputation, while at the same time the parameters that need to be monitored are crucial in determining some most essential aspects of change in our environment and climate, then shouldn't the shoe be on the other foot instead? Thus it is more important to change an obsolete ideal of science; scientists must be convinced of the relevance of a question that has to be "monitored". Monitoring is important to establish a base-line against which important environmental and climatic changes may be measured.

Scientists ought to play a more active role in rectifying the situation, so that monitoring becomes acceptable. In CCAMLR, unfortunately, Barnes noted, scientists have been lax in getting monitoring upgraded to the degree needed. If the NSF refuses to get involved in "directed research", then scientists should do something about changing the low level of consciousness, because the situation is urgent. As it is now Congress in the US, with pressure from environmentalist lobbies, have gotten NOAC to put money into monitoring in order to fill the gap. Why have we not invested more in "directed research", Barnes queried. In the case of setting "krill caps" there is a great need for such research; because of the lack of it we do not have the data needed to provide the knowledge base for rational decisions on this controversial issue, where some countries would like to see no ceilings at all.

And what about the Antarctic component in IGBP - is that not related to monitoring? Whatever it be, environmentalists want to see more of this kind of research.

One of the scientists countered that here it was a case of activities that are and are not monitoring. It all depends upon the context and the institutional motives how one defines things. The point is we don't want politicians and the media to come along and say measure this and that.

PART III
Is Science in Antarctica facing the Prospects of Increasing Bureaucratization?

CHAPTER 5

The Place of Regulation in Relationship to Science

Science Incorporated

In his review of the history of science in relationship to the ATS, Olav Orheim pointed to the increasingly diminished role of SCAR, while bureaucratic structures have

> *Olav Orheim* takes a leading position in Norwegian polar science. His concern is the maintenance of quality science, and he is well known for his sharp critique of both bureaucrats and environmentalist lobbies. In his presentation Orheim concentrated on three questions: science in the ATS, the Environmental Protocol and claims regarding the fragility - yes or no - of Antarctic ecosystems

grown, especially in connection with the introduction of conventions that have been adopted to regulate natural resources and the environment. Regarding the new Environmental Protocol, he noted that the introduction of environmental impact assessment (EIA) is not all bad; however he insisted that science needs free hands. Finally Orheim reiterated his well known view concerning the fragility of the Antarctic ecosystems, that this is a misconception; many of the ecosystems are in fact quite robust and one should not worry about incidental mortality of penguins and seals in connection with, for example, seismic surveys using explosives.

In the past monitoring activities in the modern sense did not exist. With the advent of regulatory mechanisms within the ATS the number of such activities has increased, first slowly and now dramatically. At the same time this has been part of a development that involves a reduction of the importance of SCAR and scientists generally.

The introduction of various conventions and the like can give an indication of this declining role of science and SCAR within the ATS.

When the Seals Convention arrived on the scene, SCAR's advisory role was still unchallenged. No specific advisory committee was set up; rather SCAR served the new Convention with whatever specialist knowledge that was required.

Map of Antarctica showing the research stations

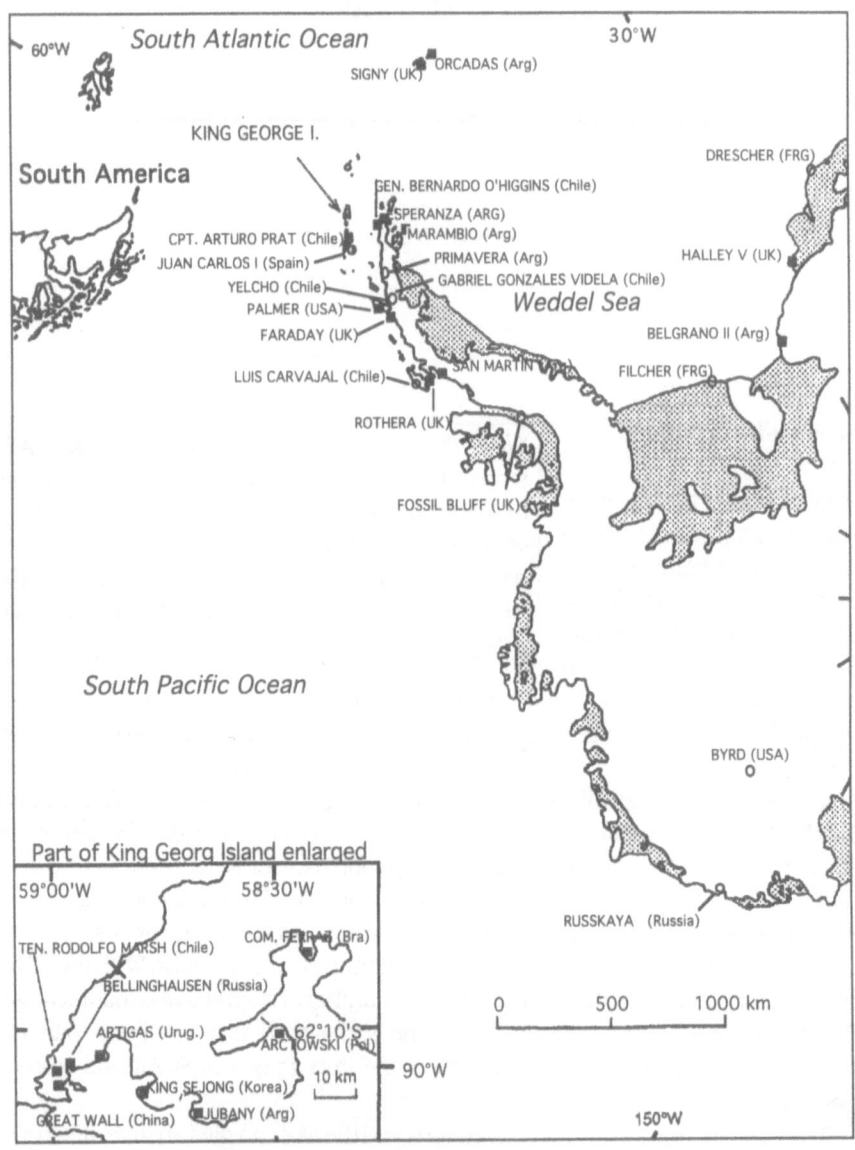

For exact location of research stations see appendix III

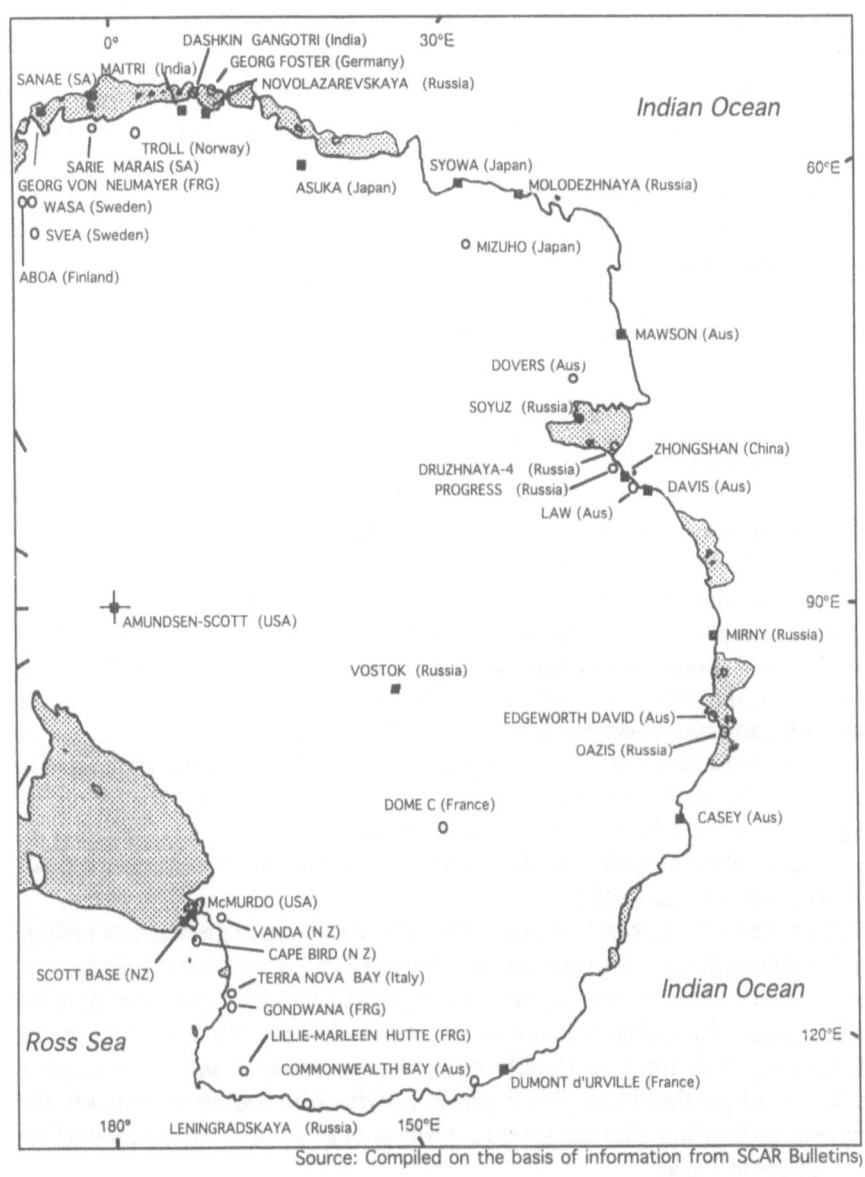

0°   DASHKIN GANGOTRI (India)   30°E
      GEORG FOSTER (Germany)
MAITRI (India)
SANAE (SA)   NOVOLAZAREVSKAYA (Russia)

Indian Ocean

TROLL (Norway)
SARIE MARAIS (SA)   SYOWA (Japan)   60°E
GEORG VON NEUMAYER (FRG)   ASUKA (Japan)   MOLODEZHNAYA (Russia)
WASA (Sweden)
SVEA (Sweden)

ABOA (Finland)   MIZUHO (Japan)

                                                        MAWSON (Aus)

                              DOVERS (Aus)

                        SOYUZ (Russia)
                                                ZHONGSHAN (China)
              DRUZHNAYA-4   (Russia)
              PROGRESS   (Russia)   DAVIS (Aus)

                              LAW (Aus)

                                                              90°E
AMUNDSEN-SCOTT (USA)   MIRNY (Russia)

              VOSTOK (Russia)
                                          EDGEWORTH DAVID (Aus)

                                          OAZIS (Russia)

                        DOME C (France)   CASEY (Aus)

                McMURDO (USA)
                    VANDA (N Z)
                    CAPE BIRD (N Z)
SCOTT BASE (NZ)   TERRA NOVA BAY (Italy)   *Indian Ocean*

                    GONDWANA (FRG)
                    LILLIE-MARLEEN HUTTE (FRG)   120°E
Ross Sea
                    COMMONWEALTH BAY (Aus)
                                          DUMONT d'URVILLE (France)
      180°   LENINGRADSKAYA   (Russia)   150°E

Source: Compiled on the basis of information from SCAR Bulletins)

■ Year around operation
○ Austral summer only

With the adoption of CCAMLR a special scientific advisory committee was set up, and SCAR consequently was given a lesser role than it had before; at the same time environmental and resource monitoring programs were set up, affecting the nature of science.

With CRAMRA this trend continued, since the convention made provisions for a special advisory committee of specialists.

After CRAMRA, which did not make it, we have the Environmental Protocol. Here the bench-mark for SCAR's role has sunk even lower. At the same time we have a much more complex system of regulations, with an interplay between different elements, and further pressure in the direction of environmental monitoring and other applied activities. Basic research workers have therewith been given a lower status in the context of the ATS. If there is any growth area for science, it is in the field of legal science and international relations. The protocol is an excellent stimulus for the production of papers in these fields, since there are a whole host of problems with interpreting the protocol, and there will be national variations. Enforcement is also a delicate and complex issue.

The Future of SCAR

So what are we to make of this general trend? Should the ATS make better use of SCAR, or should SCAR take the consequences of the fact that developments are in fact moving in the exact opposite direction? The role of SCAR for basic science is still there, but in relationship to the instruments of the ATS it is becoming more and more of an anomaly. SCAR's budget is also tiny - in total it corresponds to about 2% of the cost of constructing the new German Neumayer station; and the ATCP meetings spend as much on translations services at their biennial meetings.

Orheim's conclusion was that SCAR should pull itself out of the bureaucratic morass of politics and science and concentrate on its primary task, that of representing the interests of basic science, quality control, and speaking on behalf of its own scientific lobby. Depoliticize SCAR and let it see to its own interests and those of science, international science.

Of course there is an irony here, since this depoliticization in the sense of pulling out of the official political and bureaucratic structures at the same time constitutes a consolidation of SCAR' s role as a political non-government actor in its own right on that same arena. This Orheim did not say, but it would be one of the implications of SCAR untying itself from its double bind; depoliticization implies an implicit politicization of another kind. This is similar to what one sees in other fields, for example in connection with the tortuous process of building a Super Conducting Super Collider in Texas.

In one sense SCAR was a product of Big Science, but in its organisation and attitude it followed the ideal of little science. Now the character of Big Science has also changed, and SCAR is having to catch up and develop a clearer profile. This is essentially the question emerging from Orheim's position: how to do this. It confirms the fact that science policy has to be informed by a perspective that looks at both the epistemology and politics of science in one and the same vision.

The Environmental Protocol and Impact Assessment

With regard to the new Environmental Protocol, Orheim suggested that some of its text has been a response of politicians under pressure from environmentalist groups who have succeeded in manipulating the media, with the result that some of the rules and regulations that have been written for science are beyond what is reasonable. They are also beyond what we have knowledge for today.

Had the same regulations been applied in the past, many a project or expedition would have had to fold up and turn around and go home. Thus the Germans originally went to the Weddell Sea area to set up a research station. Difficult conditions prevented this, and instead the expedition moved along the coast further north and ended up launching a station in its present position. With the new protocol in place they would have had an EIA for the Filchner-Ronne Ice Shelf but not for the new area. Lacking such a document, the question is if they would have been able to improvise as they so successfully did; science needs free hands and cannot be tied up in a net of bureaucratic rules.

On the other hand the EIA has important positive effects. For one it forces us to do better planning. Secondly it forces us to engage more seriously in international cooperative efforts.

The protocol says a lot of things that imply more monitoring, which if followed will require a lot of scientific energy, tying it to doing more routine measurements, opportunistically to meet the new requirements and political pressures behind them.

Environmental lobbyists have been very single-issue oriented. They have held up pictures of bleeding penguins and the like, portraying mankind as the evil force in Antarctica. If we look more closely we see the Darwinistic struggle for survival where crabeater seals are over-populated and a million die annually of natural causes, or where leopard seals munch on adélie penguins, we see that nature itself is much crueler than man in Antarctica. In this perspective the loss of 20 seals blown up in connection with seismic tectonic surveys is nothing to get upset about. Environmentalists should consider the wider picture and get their proportions right. As it is they have unfairly played on peoples' ignorance and feelings of pity, and used the mass media cleverly to gain visibility for their cause. In the course of this they have painted one-sided pictures that have had a negative effect on science, both in terms of public opinion and by virtue of the pressure it has put on politicians and bureaucrats who have been compelled on the basis of partial ignorance of the real situation, to introduce regulations that go far beyond what makes sense.

The notion that the ecosystem is fragile is also a fallacy. Antarctic ecosystems are in fact very robust, since they have evolved under extreme conditions. The animals that live down there have efficient group strategies that allow them to survive as a colony, even if individuals die on the way.

Finally for those who have not been in Antarctica it is perhaps hard to understand what vast expanses of wilderness that exist there. Environmental impact from research stations and high tech projects like ice core drilling operations can never be more than very local. If one traverses the continent one will have to go thousands of kilometres before one possibly meets other human beings from some other scientific expedition. It is true that there are many stations on one small island; but to constantly harp on this fact is to distract from the fact that there are all too few research stations on the Antarctic mainland. There we need more, not fewer stations, agreed that they should be sited in accordance with scientific criteria in order to yield the maximum value in terms of science.

CHAPTER 6

The Place of Science in an Environmentally Regulated Continent

The Distrust between Scientists and Environmentalists

In his presentation James Barnes referred to the distrust between some scientists and environmentalists. In his view this had to a large extent been created by the minerals convention, which polarized relations between the two groups. For their part the environmental groups have done their best, to

James N. Barnes has been active in Antarctic affairs many years. He has been instrumental in the drafting of various documents that Friends of the Earth and ASOC have at various times presented to scientists, administrators, lawyers and diplomats at ATCP meetings as informal input to national delegations. Perhaps more than anyone else in the international environmentalist movements he has served the role of adviser from the alternative non-governmental context.

educate Congress on matters of science. It is unfortunate that scientists themselves have not been more active in this respect. Of course the position of the environmentalists is to press for more "directed research", which is something that scientists may not like. However there should be room for dialogue and mutual understanding. It is good that CRAMRA is gone; we are better without it — the preconditions for dialogue and mutual understanding are better now.

In many ways scientists have themselves to blame for the bad press. Perhaps some environmentalists individually have been unfair in their critique of scientists, but why have scientists themselves been so reticent in their relationships with the public and with politicians? SCAR should have been much more active in seeking funds and arguing their own case. In this respect the scientific community maybe can learn something from the environmentalists. It is simply a question of the scientific community getting their act together rather than badmouthing environmentalists.

The new situation with the protocol opens possibilities for scientists and environmentalists to work together against bureaucratic inertia and in obtaining more

Antarctic Peninsula and King George Island

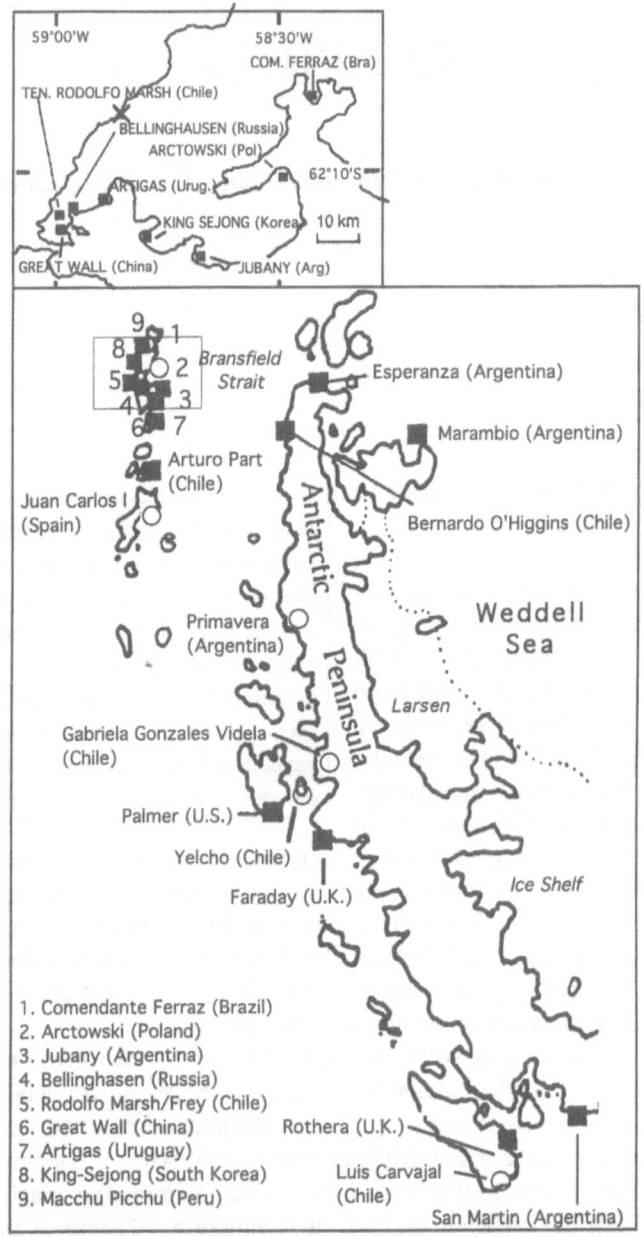

1. Comendante Ferraz (Brazil)
2. Arctowski (Poland)
3. Jubany (Argentina)
4. Bellinghasen (Russia)
5. Rodolfo Marsh/Frey (Chile)
6. Great Wall (China)
7. Artigas (Uruguay)
8. King-Sejong (South Korea)
9. Macchu Picchu (Peru)

The greatest intensity of research stations exists on King George Island, just off the Antarctic peninsula, and relatively easily accessible from South America funding for science. The new situation should also facilitate a greater degree of sharing of logistics, and SCAR should encourage governments to think about this. There is no way getting around the fact that the siting of research stations on King George Island is the result of choice on the basis of short term economic and political interests.

Here too scientists and environmentalists have converging interests, in seeing to it that new stations in future are sited on the basis of scientific criteria and combating political expediency and pragmatism.

International Research Stations

There has been some discussion concerning international bases. Here is an area where strong scientific leadership is required. The scientific community needs to be more articulate on this and state their case. Congressional committees in Washington are certainly open to arguments and would like to hear more on such possibilities (Senator Gore, for example). If scientists state their case they will find ready partners in environmentalists to support them in their arguments for more funding for science. For their part the environmentalists see two areas as particularly important. One is research related to global programs, like the IGBP; the other is research that can provide a better basis for decision-making with regard to marine resource management. Setting krill caps is a problem. Here much more scientific information is needed, with investigations focused on a fine scale study of different regions in the southern ocean.

Unfortunately in the US, the NSF does nothing to help us make the case for this kind of research on krill and its related ecosystem. Here leadership must come from the Antarctic science community.

The Environmental Protocol and EIA's

With regard to the Environmental Protocol, Jim Barnes had difficulty seeing what all the alarm is about. Where is all the worry coming from, he asked. Remember it was SCAR that came out with the first proposals on EIA. Moreover it is the logistics that supports the science that is the culprit when it comes to pollution; it is not the science itself that we point the finger at.

Barnes maintained that the fear of over-regulation is a red herring. The protocol makes provisions for "emergency situations", and such provisions may well be invoked in the case Olav Orheim described with regard to the original placement of the German Neumayer station.

Also, alternative channels of scientific advice in the ATS are OK, but hopefully SCAR will continue to function in this regard too - it is SCAR that has an authoritative scientific position. It would be a pity if SCAR pulled back; the conclusion should be the opposite, to articulate a stronger leadership profile, proactive rather than reactive.

SCAR's Leadership Role Scrutinized

In the discussion following these presentations, Bruce Davis agreed with the diagnosis that SCAR has been rather passive and that the Antarctic science community should come forward with a clearer message to politicians. After all scientists also constitute a single issue interest group, just like the environmentalists. We live in the era of group politics, and this calls for conscious leadership in the community in question. Within science itself the Antarctic people also only make up a very special group that has to compete with other groups for scientific funding. Some of this competition takes place through the National Academies of Science and the research councils.

Jarl-Ove Strömberg reminded those present that SCAR is not a body that implements decisions on an international arena. It is only a body that works through the National Academies and the like, which means that variations in national policies make it improbable that SCAR can act in the united way one might like to see; this is especially the case when it comes to issues involving politically sensitive questions. That is the way ICSU bodies work. Thus it is difficult for SCAR to assume the leadership role that is being asked for here.

The Role of the NGOs

Several of the scientists at the meeting were pleased with Jim Barnes' statement that the NGOs are willing to work with SCAR in a united effort that would be to the benefit for both science and the environmental question. Nigel Bonner for his part also noted the different opinions within the scientific community. "I am looked upon as a rabid green by some of my colleagues in SCAR", he said, and went on to indicate that IUCN (Paul Dingwall, NZ) and the CNPPA could offer more to SCAR (the latter is the advisory group on national parks).

Some of the discussion came to revolve around the behaviour of the environmentalists. Not everyone accepted Barnes' description of the situation. Tension between the two camps has developed because of some environmentalist's manipulative use of the media and sensation-seeking methods. Scientists also feel misunderstood, snubbed, frustrated, hurt and angry. It is not so strange that they have reacted as they did to the Environmental Protocol in some cases, with scepticism. "We are concerned about the future". This summed up some of the sentiments of scientists. It is clear that the two camps have different perspectives of how the objective of environmental protection should be achieved, even if there is general agreement about the objectives as such. One has to be realistic; the handling of toxic wastes may in some cases require services that do not exist, for example.

Different National Styles and Scientific Cultures

Nigel Bonner agreed with Olav Orheim's claim that the ecosystem in Antarctica is robust. He disagreed however about the claim that there are too many stations on King George Island. If one considers one place, Maxwell Bay, perhaps yes, but in general no, there is plenty of room on the island, and the stations do not make any significant impact on the environment. Orheim countered by explaining that this is not what he meant; indeed the impact on King George Island is not large, but it is not all good science, since we get a lot of duplication with stations so close to each other.

Another question that came up was Jim Barnes' plea for a more far-reaching internationalisation of Antarctic research activities. Here one could notice a difference in approaches which may reflect different national styles. The British were quite resistant to the idea, equating international stations with disorganisation. Barriers of culture are an important limitation. Also an international base would create even more bureaucracy, a "horrible thought". Multilateral cooperation in expeditions was however quite acceptable.

The Dutch and Scandinavian participants were more favourable, but also realized that there are significant cultural barriers and differences in mentalities that can make it unrealistic to think in terms of a wholly international base. Scandinavian experience reveals that multilateral cooperation in the logistics phase can be successfully implemented, and reference was made to the "Nordic" model of rotating logistics responsibility between countries - Sweden, Finland, Norway. Anders Karlqvist agreed that the prospects for international cooperation were never better, but that they were not as rosy as Jim Barnes had painted them. In practice there are many complications. It works best when someone has the initiative and invites someone else along. Many large scale international projects fail because of too many rival wills. An example mentioned is the deep ice core drilling on the top of the Greenland ice sheet, where there now are two holes near each other, one European and one American. One can certainly ask if this duplication was necessary. However a variety of factors brought about by differences in funding systems, bureaucracies and politics led to the situation as it stands.

The Nordic model, it was suggested by some, is possible because there is a greater affinity of mentalities and sociocultural factors than there would be in the case of joint ventures involving countries with dissimilar backgrounds and traditions.

Greenpeace, Safety and the Role of the NGOs

Environmentalists were also given some praise. Greenpeace, it was noted, had done an excellent job on its inspection cruises around various research stations. This has had a positive effect in the Antarctic science community, which needed some external prodding to start cleaning up its act "down there". This is the kind of work Greenpeace should continue to do in future, to play a kind of watchdog role, since the Treaty cannot or will not do it.

The Greenpeace base at Cape Evans is now being removed. It has served its purpose, which has not been science, but rather obtaining visibility for this organisation and its view on Antarctic affairs. Thus environmentalists should not come and tell scientists about the siting of stations after political criteria. The Greenpeace base was placed in Antarctica for political purposes, an act quite comparable with Equador's base.

Olav Orheim added to this that the international environmentalist lobby was good at playing the power game, and that the Greenpeace base actually was part of a campaign to get more money from a willing public.

Paul-Christian Rieber returned to the question of the scientific community, agreeing with those who emphasized the need for it to act more forcefully. Accidents also cause bad publicity. Here scientists have also been rather lax; they should have taken a more active role in seeing to it that the equipment used is right for the task, and that vessels operate at a high level of safety. This of course also goes for tourist operations - then catastrophes like the one with the New Zealand airplane or the Bahia Paraiso could be avoided. Safety is an important criterion, as already discussed by Anders Karlqvist.

Jim Barnes replied to several of the points that came up. First he clarified the purpose of the Greenpeace operation in Antarctica, saying that the NGOs did not set up a base there in order to influence public opinion to get money, as Olav Orheim claimed. Rather it was to gain a first hand experience of Antarctica, its environment and conditions in order to obtain a better base of knowledge from which to evaluate different countries' activities and argue for an alternative approach to Antarctic affairs on the basis of this knowledge. "We wanted to get experience and more direct information", Barnes said, and pointed out that until 1983/84 the ATCM's were a closed forum, which made it difficult for outsiders to gain information on what was going on down in Antarctica.

Secondly, Barnes meant that if the ATS wanted to set up an environmental police force this would be a good thing. The NGOs do not fancy themselves as self-appointed environmental police, but have been driven into this role because of the laxness of the ATS. If governments do not do it, then Greenpeace and ASOC will have no choice but to continue their watchdog operations in one or another form.

Thirdly, as for the NGOs track record, it is not a bad one. A lot of documents have been produced, with constructive proposals and factual information which has been broadly disseminated. This has also been helpful for the newer nations coming into the ATS. The media have tended to capitalize on sensation and distort things, and some Greenpeace groups admittedly have also been a bit flamboyant; but in the main the NGOs have played a very constructive role. The inspections tour of many stations in order to determine environmental impact and what waste management methods are implemented in various locations has been appreciated by many scientists, who have seen the reports as constructive and fair.

The Minerals Issue

In connection with the minerals question there has been a lot of polarization of views, but within the conservationist community both liberal and conservative elements have been united in their position that the minerals connection should be stopped. Interchanges with people in the scientific community have sometimes been sharp, but ultimately we have one and the same interest at heart, Barnes said, to preserve Antarctica for scientific purposes and aesthetic values. This has not been sufficiently understood; perhaps now the minerals business has been put aside mutual understanding will increase.

In the case of the French airstrip at Dumont d'Urville, the NGOs had information that this was part of a plan in which considerable attention in research was being devoted to minerals research. Even in France the scientific community was divided, and there was debate at Cabinet level amongst members of the French government. Thus scientists and politicians were both divided on the airstrip issue. Of course this situation was important for the tactics of the NGOs, and in the course of polarization things may have gotten a bit too simplified, turning into a debate on people vs. penguins.

Nigel Bonner, who was involved in a review of this case at the time within the French governmental system (a Comité de Sages providing input to the Territoire des Terres Australes Antarctiques Françaises -TAAF- an autonomous unit under the Ministry of Overseas Territories) was able to give some further background information. The question as it appeared from his vantage point was one of ensuring the safety of people, and providing better conditions for good science, among other by giving scientists more time in the field by the more rapid route of access by air instead of ships. This was an important point for scientists. Also the French plan to set up the new Dome C station played in, since here it was even more important with earlier access. The d'Urville airstrip will function as a feeder for traffic further into the interior, where interesting astronomical work and ice studies are being done. Bonner said that at the time he and also El Sayed the latter as convenor of the group of Specialists on Southern Ocean Ecosystems, accepted this argumentation, and he added somewhat self-critically, "We didn't explore the alternatives in a proper way".

PART IV
Orientational Shifts in Antarctic Research Agendas

CHAPTER 7

Focussing an Antarctic Research Program - the Australian Experience

The Australian case is interesting in that we have a country that has played a leading role in the move to push for a comprehensive environmental protection regime within the ATS. This was after first having endorsed the negotiations leading up to the Wellington convention on minerals.

Bruce Davis is Deputy Director of the Institute of Antarctic and Southern Ocean Studies (IASOS) at the University of Tasmania, which is in the process of reorganization and expansion. He combines an engineering background with expertise in the social sciences, and has been involved in several evaluations and consultancies regarding science and technological development. Dr. Davis has written widely on policy matters.

Bruce Davis highlighted some of the events that led up to a restructuring of Australian research on Antarctica, and he described the present situation when environmental management in Antarctica has become the overriding political issue.

An important question has been how to review and evaluate scientific performance. Does one measure the numbers of scientific papers Antarctic researchers put out each year, or something else?

Another question concerns the optimal form for organizing the production of knowledge in this field. Should this be inhouse within a single national agency as it is now - the Antarctic Division - or should the main responsibility lay in the academic institutions? Presently it appears that Australia is moving from a single agency model to one of a mixture which would give academic research a more prominent role. This would in turn, it might be argued, facilitate greater attention to peer review and quality control. At the same time management of the Antarctic environment is being emphasized, which tends to pull science into the direction of a socially mandated service function linked to monitoring institutions.

Historical Background

Australia was already involved in Antarctica during the heroic age, particularly with the exploits of Douglas Mawson. These belong to the annals of the heroic age. Funding in those days was based on public donations and private sponsorship. In 1928, when Mawson organized the British-Australian-New Zealand Antarctic Research Expedition (BANZARE), the political dimension was most prominent, with government giving some input. During this expedition 1929-31 a lot of new territory was mapped in the little-known Indian Ocean sector, which laid a solid basis for what was later to be proclaimed as Australian Antarctic Territory.

Mawson's first expedition 1911-14 yielded scientific reports that are collected in 22 volumes, dealing with meteorology, geomagnetism, geology, biology, oceanography, and the aurora. The BANZARE reports comprise 10 volumes or more. For the most part, however, the resulting science during that period up to 1945 was not well structured. Davis referred to this period as one of idiosyncratic individualism.

In the post World War II history of Australia's Antarctic involvement, two further eras may be distinguished, one from 1945 to 1990, and the other the post-CRAMRA era that has just begun. The first is the era of increasing attention to science, the second one of environmental management, just beginning. From 1945 onward it was a question of gaining a better foothold on the continent and entrenching it; now it is a question of Australia taking the lead in making good the reorientation towards environment.

In 1947 the Australian National Antarctic Research Expeditions (ANARE) was created, and a special Antarctic Division was established under the Department of External Affairs in 1949. From 1948 to 1969 activities largely reflected the drive and interest of individual sciences, as well as the desire to have a good spread of scientific disciplines and specialties in Antarctica. Stations were established on Heard Island and Macquarie Island, and in 1956 the Mawson station was set up on the mainland, in good time for participation in the IGY. Gradually Australia increased its effort to maintaining three continental Antarctic stations, which some claim is not very much considering that these are to match a political reality where 42% of the continental area is claimed. Until 1969 research continued at a steady pace, with none of the visibility it has received since then. Coastal exploration, inland reconnaissance and station building programs have absorbed much of the funding, but now Australia had programs in earth sciences, glaciology, life sciences, and upper atmospheric physics.

In a second phase of the science era, i.e., 1970-1979, there was a considerable amount of rethinking and reevaluation, as Antarctic affairs came into the limelight. At the political level demands were made to strengthen Australia's Antarctic presence. In practice this was to occur through the symbolics of a heavy station building program, which later drew criticism.

A new planning committee, established in 1973, the year of the oil crisis, exerted some influence. This was the Advisory Committee on Antarctic Programs (ACAP), which proposed several new initiatives, among them a program to rebuild the

Antarctic stations, a marine science program, and a plan to build an Australian research vessel for Antarctic programs.

In retrospect the politics of all this appears to be in line with a move towards a principle of effective occupation through science. This was criticized by the next advisory body, the Antarctic Research Policy Advisory Committee, which was established in 1979, marking a new turning point.

Changing Organizational Structures and Current Priorities

ARPAC put greater emphasis on longer term scientific content, and the rebuilding program was criticized for not taking into account the principal strategic objectives of science, nor the need for better transport and enhanced logistics. Because of its critical stance, ARPAC came into conflict with the government and was disbanded in 1984. The critique launched against ARPAC was that it became too much involved with logistics and started to interfere with the Antarctic Division's day to day management of research and other activities. A new committee was created, this time with the acronym ASAC, short for Antarctic Scientific Advisory Committee. Its mandate is science policy, and to keep out of logistics. It advises the Ministry of Environment broadly. Tension has however also developed here, between the Antarctic Division and ASAC's perception of this agency's work and how science should be done. The period 1985 to 1991 is being punctuated by a new review of Australia's policy regarding Antarctica and related research programs. The result of this review will be in hand late 1991 or early 1992. The scientific dimension is having to be re-focussed in consideration of the political reorientation which sets up environmental management as a central issue. The political decision has been made, but now corresponding institutional rearrangements have to be put in place, both at the level of policy implementation and at the level of research performance.

Presently Australia has seven priority program areas. These are:

* unique Antarctic sciences
* earth sciences
* weather and climate
* CCAMLR
* technology and support
* environmental management
* social sciences

Economic constraints in the country have prompted a need for greater accountability and steering in science. This is also being felt in the field of Antarctic and southern ocean studies. With the prospects of an increasing focus on environmental management activities, both bureaucrats and scientists in the Antarctic research area are asking for more funding for this particular purpose, so that the new task will not cut into the science budgets.

Political Symbolics and Research Efforts

There is a lot of political symbolics at stake here. Australia, together with France, abandoned CRAMRA in favour of a more radical and comprehensive environmental line. Now, so the argument goes, Australia will have to back up this posture with sufficient funding in order to have something to show the world. Otherwise the shift to environmentalism may be interpreted as yet another expression of opportunistic self interest related to entrenching Australian influence in Antarctica, the 42% of the continent that it once claimed.

Thus a tension builds up between environmentalist interests which have been strongly pushed at the political level, and scientific interests which were the currency of Antarctic politics in the previous era (1945-1990). Unless new funding for environmental management purposes is forthcoming, this tension will increase.

The Australian effort in science is after all a limited one when we look at it in terms of funding. Only about Aus$ 8 million of a total Antarctic budget of Aus$ 70 million goes to research. ($Aus = 0.8 $US). The cost of bureaucracy and also that of transportation and logistics has gone up, while the science budget has remained about the same as it was. Most of the money goes to Antarctic Division. The universities get only about Aus$ 1/2 million (ASAC grants program). In addition to this the universities of course can use ARC grants, and logistics is provided for out of the Antarctic Division budget.

The debate is now, to what extent resources should be concentrated to centres of excellence, or to what extent the concentration on Antarctic Division should continue. It is clear that the universities should be given a more prominent role, but how? Also, in view of the emphasis on environmental management, the promises of new funding for science is perhaps more a question of high level rhetoric than reality.

The NGOs for their part have been active in pressing for more money for Antarctic environmental conservationist interests. This would also imply a new spectrum of research priorities, in fields such as environmental impact assessment, waste management systems and nature conservation in protected areas. Australia it is said has to set an example in environmental management of a standard that other nations may emulate.

Scientists for their part, here as in other countries, fear that some of this reorientation may be at the cost of their traditional science programs, and that scientific activities may become too restricted in various ways.

According to Bruce Davis, it is clear that new emphasis and priorities arising from environmental conservation measures are becoming institutionalized in Australia. However this also necessitates reeducation of attitudes and values, as much as some reordering of scientific programs. Furthermore, there is the debate about where to put the resources. Some say that research carried out by Antarctic Division is too expensive and dependent on interests of individuals rather than national priorities. The counterargument is that some of the research projects funded outside the Division are subject to voluntarism and instability, since they are built up around postgraduate students, with limited focus and short term duration. The universities do not like to send out senior researchers to Antarctica, since it takes a lot of time, much of which is in transit and improductive. Antarctic Division allows for longer term programs and greater continuity in research.

The verdict on these and other issues will be in hand soon, as the review committee comes with its recommendations.

Discussion

The discussion picked up on several points. One was the question of the heavy station building program, another on the relative virtues of the central institute model contra the university research model. With regard to the latter it was pointed out how the universities never can give you the presence Australia needs for political reasons as a claimant country. Someone wondered if there were any scientists in Antarctic Division, or if it was overgrown with bureaucracy. The thrust of the remark was to urge Australia to build up scientific capacity in Antarctic Division and replace individualism with an effective central divisional control and leadership. In this way you can build up the kind of balance between basic research and science for environmental management purposes you want, and also get a better integration of the university-based research efforts into existing programs. What is needed is strategically planned programs.

Bruce Davis agreed, but insisted that there was a difficulty in pursuing this line of development: "To get Antarctic Division together is not easy".

In this connection the logistic aspect also came up. Australia has built a new vessel at great cost. Couldn't logistics costs be reduced by chartering vessels instead, even in future, while reducing bureaucracy? If you reduced outlays for logistics and bureaucracy to an extent of 20% of the total budget this would free an equal amount for research, which means that the science budget could be doubled. What is stopping you from doing something along these lines; within the private sector this would be a self evident approach to increase cost effectiveness.

Another comment went in the opposite direction, suggesting that Australia really needed two vessels, one for logistics and one for research, especially in marine science. A clear-cut division of labour is more effective than trying to combine the two different functions - logistics and research - in one and the same vessel, because that leads to a lot of collisions in everyday life on board.

Davis pointed out that the decision to build a vessel was a political one, indicating that it was not first and foremost founded on sound scientific criteria.

CHAPTER 8

Environmentally Driven Research - is it Different?

Barry Heywood began his presentation by distinguishing three possible interpretations of the title he had been given. Environmentally driven research, accordingly, can mean any one of three things:

> *Barry Heywood* has over 30 years experience as a fresh water and marine scientist in Antarctica. He is currently the Deputy-Director of the British Antarctic Survey with special responsibility for maintaining a first class science program.

- environmentally or politically driven research;
- environmentally benign research; or,
- research about the environment.

The first of these has to do with monitoring. Currently there is a discussion regarding the differences between scientific monitoring and environmental monitoring. The more the pressure is in the direction of the second of these, the more we shall see the best brains in research on the Antarctic lost to other fields. This is because scientists have a predilection for basic research. Science as the pursuit of truth is an exciting and open-ended affair. Therefore, in fact, it is a misnomer to speak even of "Antarctic science". This terminology came in when so-called "areal studies" were in vogue. However, the prefix "Antarctic" is not needed; indeed, it is objectionable, because what is done is "science" per se. It happens to be in the Antarctic, but that in itself cannot be a defining element for the research done, Heywood stated. The primary factor is that the activity be science-driven. This is what makes it worthwhile for the scientists concerned.

With regard to the second category, environmentally benign research, this is no problem in Antarctica. Scientists need a clean baseline for their observations. Thus there is a natural inclination to keep the environment around their observation platforms clear. In analogy, who wants to use dirty test tubes to perform sensitive experiments?

Finally, the third category - research about the environment - is no problem either, as long as there is something in it for science. If it is scientifically interesting, geared for example to climatological studies, or biological programs, such research is welcome. Time, costs and people are a limited resource, and thus one has to be selective in the targets focused. Scientists will want the targets to be able to give a good scientific return.

The Structure of the British Effort

The principle is that one should in the Antarctic do only those things that are best done down there, for example investigating the properties and dynamics of the ice cap. Studies that are equally well done elsewhere can be excluded, because the cost-benefit equation in Antarctic will be much higher.

Another principle is that the research done down there must follow the same high standards as research done anywhere else. Therefore it is important to subject Antarctic activities to stringent peer review criteria. Peer review by internationally established scientists will also reinforce links with disciplinary research that has not had Antarctic aspects.

Barry Heywood's conclusion regarding the Madrid protocol was that it is welcomed by scientific communities, provided the implementation will help to promote and not hinder good science. This is a point to which the discussion came back to later.

After the foregoing presentation of the issue and his own reading of it, Heywood reviewed some highlights of the BAS, its structure and programs. Generally BAS covers the main types of research encoded in SCAR; basic division of disciplinary and specialty areas. The effort is structured in accordance with five themes:

* Patterns and Change in the Physical Environment of Antarctica
* Geological Evolution of West Antarctica
* Dynamics of Antarctic Terrestrial and Freshwater Systems
* Structure and Dynamics of the Southern Ocean Ecosystem
* Physics of Solar-Terrestrial Phenomena from Antarctica

In addition to this there are two further research programmes. "Humans in Isolated Polar Communities", and "Antarctic Geographical Information and Mapping". The latter has been strongly stimulated by modern computing, which allows for coordination and improvement of data bases, with input from satellite imagery and digital terrain models. The effort involves wide-ranging cooperation with scientists in several other countries.

Each of the programs is further subdivided into many different projects. These are the outcome of successful applicants receiving funding. Applications for project

The Mezozoic Supercontinent of Gondwanaland

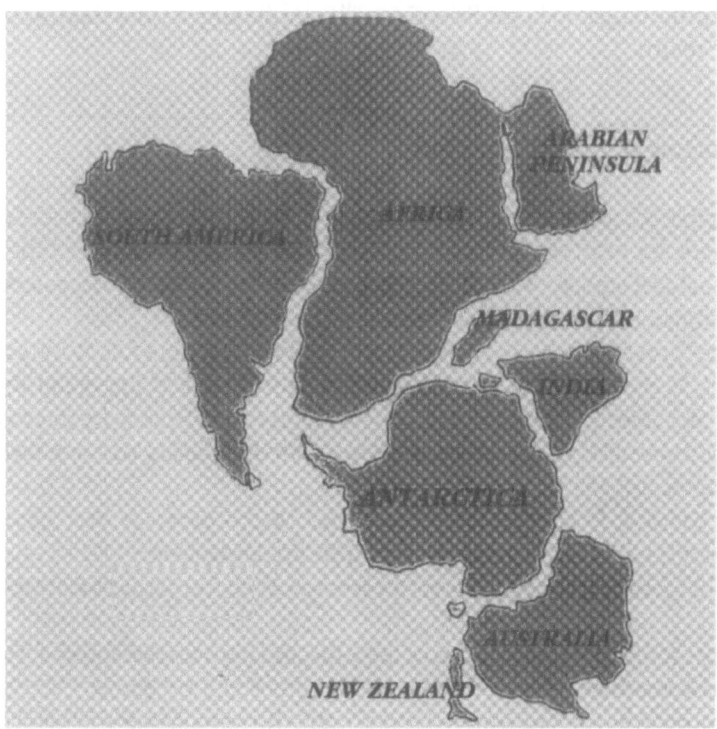

This super continent comprised the land masses shown, and it began to break up about 180 million years ago when the different parts began to "drift" apart towards their present configuration. The mountains along the Antarctic peninsula extended under water into the Andes of South America, while the Eastern part of of Antarctica has rock formations that indicate connections with similar ranges in South Africa and Australia, respectively. Thus it has been assumed that there must also be similarities in mineral resources in certain areas of Antarctica compared to those found in the mountains on the other continents. Because of its strategic location in Gondwanaland, Antarctica holds the key to many puzzles being worked upon in geological sciences, paleoclimatology, glaciology and a variety of other fields.

grants are evaluated on the basis of novelty, scientific merit and degree of relevance to the theme of a given program area. Every five years a peer review is undertaken of the programs as such. This process usually involves ten or more world class scientists, who look at the quality of the results obtained within the various programs.

Projects are also subject to review of the environmental impact they may have; applicants are asked to supply such information, and BAS has a special Environmental Officer who goes through all the project applications. Apart from general environmental impact, special attention is paid to measures for waste disposal, to see if these satisfy requirements.

Funding

In all, we were told, some £ 22 million is annually expended for research in the Antarctic through the BAS. Another £ 1 million is funded by universities, and the Scott Polar Institute has a budget of one half a million pounds annually. In addition NERC has special topics funding, some of which goes into polar research. Total expenditures on science in the Antarctic are therefore in the order of £ 24 million. This is an increase in cost effectiveness compared to the beginning of the previous decade. Owing to a rapid expansion of activities in the 1980's, BAS management and the structure of programs was reorganized so as to facilitate greater cost-benefit consciousness, and activities have indeed been made more cost effective.

Over and above the recurrent funding that has been referred to, BAS has been given additional resources for a couple of capital intensive construction efforts. One of these is the Rothera airstrip, which will be in service in the Fall of 1992. BAS is currently seeking additional funding for the rebuilding and expansion of the research facility on Signy Island, to increase its capacity from 27 to 40 scientists. The airstrip for its part will enable BAS to bring in university scientists for shorter periods, thus broadening the constituency of field workers. Furthermore, the airstrip will extend the range of operations in the West Antarctic icefields.

EIA as a Management tool

Both of these major construction efforts have been subjected to a thoroughgoing EIA, with representatives of environmental organizations being invited for on site inspection visits. EIA, Barry Heywood observed, has proven to be a valuable management tool. Its ultimate effectiveness will of course depend on national procedures, and here one may expect some variations. Some countries will no doubt follow recommended procedures carefully, while others may be less apt to follow suit, either through ignorance, or due to a lack of the resources required to implement required norms. Implementation of the EIA provisions, Heywood insisted, is best left to the self regulating mechanisms of the scientific community. If one takes a more stringent approach, as New Zealand, Australia and France have asked for, to consider all activities as harmful unless proven otherwise, this might lead to some nations pulling out of the Antarctic Treaty but without leaving the Antarctic. This would create a situation that no one wants. Also, if this became the dominant interpretation it would lead to a reduction of science for other reasons, among them the scientific community's frustration in the face of what would be perceived as over-zealous control. It is important to strike a balance, and to emphasize the principle of scientific self regulation. Much can be achieved by patient counselling; what is needed is education, not legislation - counselling, not policing.

The Role of SCAR, once again

Much of the discussion revolved around the BAS format for research grant applications, which was found to be exemplary and worthy of emulation in other countries. Also the introduction of an Environmental Officer that checks all proposals with respect to environmental impact and waste management gives Britain a good lead in accommodating to the requirements of the new protocol.

Barry Heywood had observed how the implementation of EIA's will be dependent on national procedures. Jim Barnes, appreciative of BAS's mechanisms in this respect, urged that SCAR might take a leading role here by requesting all countries to use similar formats for project proposals so that a standardized approach might be developed, giving criteria both for quality control and check-lists for environmental impact. He also suggested that inspection procedures might be developed. The ATS allows for site inspections, but this is all too infrequently implemented. With the EIA's there is good reason to generate standardized check lists that might be employed for such inspections, which for that matter might be carried out jointly by groups of a multinational composition.

Bruce Davis agreed with Heywood about the importance of education rather than legislation, but thought this would take some time before the effects would be noticed. He was also less optimistic about the natural inclination of scientists to implement far-reaching EIA procedures on the basis of "self regulation" stating that "Scientists also need to be watched".

In conclusion there was some agreement that SCAR had a role to play here, for example in pushing for some kind of standardization in the implementation of EIA's and, in particular, asking for all countries to adopt one and the same kind of format for project proposals, which could be based on the British experience outlined above.

CHAPTER 9

Geoscience - Basic Research or Commercial Prospecting?

The Negative Image of Earth Scientists

Kent Larsson in his presentation went directly to the heart of the matter regarding the negative image earth scientists have come to have over the past fifteen years. In his estimation there are several factors responsible for this,

Kent Larsson is professor at the Geology Department of Lund's University in Sweden. He is also a member of the SCAR working group on Geology. Presently he is involved in a Swedish project on the geology of mountains in the Dronning Maud Land region of Antarctica to study sedimentation and basalt rocks, as well as biotic developments in the past. Kent Larsson is involved in research that is meant to contribute to a better understanding of the breakup of the super-continent Gondwanaland and has a bearing on plate tectonics. The SWEDARP 1991/92 expedition in which he participated complements earlier Swedish studies in the region - SWEDARP 87/88 and 88/89, in Heimefrontfjella and Vestfjella.

among them the earth scientists' own failure to come forward more aggressively to counter commercial pressures. Instead there has been a lot of opportunism, whereby they have let the image of economic utility become widespread, in the belief that this would facilitate greater funding. In part this has been true, but at a cost to the reputation of the earth sciences. Within the SCAR working group the public image problem has been discussed and Larsson noted how SCAR itself also has been too passive. SCAR should have come forward to formulate a code of ethics for earth scientists and stigmatize their participation in commercial ventures run by, for example, oil companies. Such a step has not been taken until recently as commercial pressures have declined in tandem with the shift to the environmental motive. Thus it is not so strange that the public image of geologists and geophysicists has been one of the "bad guys" who disguise minerals exploration under the mantle of science. If SCAR had exercized its control function that it has the mandate to exert on behalf of the basic research community, some of this misunderstanding might have been

81

Minerals in Antarctica

Coal-bearing areas

Ag - Silver	Cr - Chromium	Mn - Manganese	Pt - Platinum	U - Uranium
Au - Gold	Cu - Copper	Mo - Molybdenum	Sn - Tin	Z - Zinc
Co - Cobolt	Fe - Iron	Pb - Lead	Ti - Titanium	

With the oil crisis in the early 1970s attention was turned to the natural resource "treasury" locked up in Antarctica. Several countries began exploration to determine the potential for hydrocarbons, oil and gas, and there was also speculation concerning the metallic minerals located in the continent's four mineral provinces. The economic motive had a bearing on research agendas and sparked off negotiations over a minerals regime, which when finally drafted became defunct. Instead, on October 4th, 1991, the Environmental Protocol was adopted, putting a ban on minerals explorations for at least the next fifty years.

avoided. Now that a fifty year ban has been imposed on exploration and exploitation perhaps it will be easier to defuse emotions and correct the image that has developed. Of course the situation of the earth sciences differs from that of biology, both historically and factually. While the latter have always come in after the industrialists criticizing them, the former have acted as a vanguard for industrial interests and toned down their criticisms. Marine biological resources were exploited very early on, first seals, then whales, and now krill and fish. The exploitive activities were squarely in the hands of the industrialists, while biologists came in afterwards to assess the damage, set quotas and promote conservationist measures to restore ecosystems. Thus their image has been closely associated with environmental preservation - the "good guys". With geologists the opposite has been the case, they came in as a vanguard, collecting data that could be used for exploration, thus opening the way for commercial interests. Also when earth scientists have produced scientific reports they have often gone on to assess the commercial potential of minerals and hydrocarbons that have been identified as existing in various locations. Finally, suspicion has been further fed by various commercial exploits, such as Petrobas doing seismic work for Brazil in the Bransfield Strait or Argentinian oil companies having gone into the James-Ross Island region. Japanese and Russian seismic mapping during the past decade have also contributed, especially since these countries have been very much against publicly disclosing their findings.

New Opportunities

Presently there is an effort afoot to pool seismic data in a large international library, and even now there is considerable reluctance on the part of many parties involved to share their data. However, now this is mostly due to the fact that scientists want to "milk" their own data as much as possible before releasing them for a wider audience. This is in line with the role of publish or perish, which may be more important when the economic motive for data generation has declined. Of course a situation similar to the one during and after the oil crisis of the 1970's and OPEC's actions may emerge again in the future, whence one can never be sure that the ban on minerals exploration will be absolute. It all depends on global economic conjunctures and politics. Another contrast between biology and the earth sciences is that whereas the former yields data that will have a scientific and potential environmentalist interest, dual usage in the latter case will always potentially incorporate a commercial component. This remains true even if commercial pressures have lifted, at least for a couple of decades to come. The previous commercial motive for keeping field date confidential for long periods has therewith been removed.

At the same time there is now ample opportunity for the earth sciences to contribute to the environmentalist motive. This may somewhat offset the present lack of an economic motive to drive geology and geophysics. Thence earth scientists may now change their tarnished image by involving themselves centrally in research programs relating to global change, cooperating more widely in international efforts such as the planned trans-Antarctic traverses to do seismic work, or ice core drilling activities in various locations. These all have an important bearing on our knowledge of the past climate and the history of the Earth. SCAR's working group has also become clearer in that a code of ethics has been articulated, recommending scientists not to get involved in commercial ventures. The SCAR conference in Bremen in September 1991, with its focus on the environment and the attempt to raise the visibility of this aspect of Antarctic science, was also important in the remoulding of the public image of the geosciences.

Are there Mineral Resources and Hydrocarbons?

Kent Larsson started his presentation with a factual review of what minerals and hydrocarbon resources exist in Antarctica. This presentation led to several comments, with some participants maintaining that he gave a falsely optimistic picture of Antarctic mineral wealth. Opinions also differed as to the way such facts ought to be presented to a wider public, for fear of creating misunderstandings. Someone suggested that scientists should always tone down what is there in terms of resources when talking to a lay public. Kent Larsson however maintained that it is more important to be objective than tactical in such reviews.

By and large the picture he painted is one of many interesting formations in Antarctica, ones that display similarities with areas in Latin America and South Africa where mountain chains are known for their rich ore bearing characteristics. Iron and copper have been found in pre-Cambrian rocks right across the continent, from Wilkesland to Dronning Maud land, and especially in the Prince Charles Mountains. The Dufek Massive in the Pensacola Mountains holds molybdenum, chromium and other metals also found in South Africa. The youngest metalogenic province is the one related to the South American Andes with findings of Antarctic copper, vandium, silver and gold. The Ellsworth Mountains area, together with the Antarctic Peninsula and the Shetland Islands are also promising regions.

Having said this, it must of course be added that the lack of technological and market accessibility has mitigated against serious commercial economical interests.

When it comes to hydrocarbons, there have been no real discoveries, except minimally in the Bransfield Strait. There has been no deep sea drilling into the ocean bed crust, which would be needed to obtain more precise information. A commercially interesting find would have to promise at least a giant oil field with a potential of over two billion barrels. Despite speculation during the late 1970's such a find has not been forthcoming. Coal, for its part, is mainly concentrated in the trans-Antarctic mountains and the Prince Charles Mountains, and it is inaccessible with present-day mining techniques.

Retrospectively, it can be said that a lot of useful data has been accumulated thanks to the strong natural resource interest, which has now declined. This data is now

ISBERG, SEDDA DEN 11 NOV. 1902.   *Skiss af förf.*

Iceberg seen 11th Nov. 1902 during the Swedish Antarctic expedition led by Otto
Nordenskjöld. Sketch by S. Duse

becoming more widely available, and it is proving very valuable for geological
purposes. Today the dominant interest is scientific.

The discussion touched on several questions apart from the review of minerals
resources in the Antarctic. There were some differences of opinion regarding the
quality of the minerals present in Antarctica, e.g., iron. Also the different ages of the
Dufek Massive compared to the Rushfield Massive in South Africa, it was suggested,
makes a difference. Kent Larsson responded that there is no doubt as to the
significance of mineral findings on the Peninsula, and that in other cases the data
leaves considerable leeway for different interpretations and even controversy,
amongst experts. Asked why he himself was interested in geological research in
Antarctica, Larsson answered, curiosity, as well as questions relating to plate
tectonics, climatic variations and other matters of environmental relevance.

Several persons emphasized the global approach that is required in the study of
the Earth's history. Problems of geology, it was stated, are process oriented, and the
processes looked at are increasingly global ones.

85

Dual usage of Scientific Data

The issue of dual usage of geophysical data was also discussed. On the one hand there is a short term scientific interest in data, on the other a long term prospective commercial one. Thus geophysical research emits a double set of signals. This is unavoidable. When you speak to people in government the one dimension is put forward, while the other is more central within the scientific community or in addressing environmentalists and a broader lay public. In approaching government you need to play up utility as an argument; in the past decade it was the prospective economic utility of geophysical research that counted. This is a fact of life, and certainly geologists have used such arguments to justify their science. Today the shift is towards the environmental motive. References to better information on global change unfortunately are less powerful in the ears of the politicians. It is not a sufficient justification, even when one points to increases of 1-2 mm rises of the sea level per year. With the shift of focus to global change, geology may be at a losing end, Olav Orheim stated, and he went on to compare the situation in the 1970's, when oil companies gave the university of Bergen multi-channel seismic equipment for purely scientific work. They were convinced that this was an investment in the future, since it would stimulate the buildup of a general competence in this field within the university. This could be tapped when needed for commercial purposes. In 1977 the Norwegians went to the Antarctic to do multi-channel seismic work.

Jan Stel disagreed with Orheim that geology might be at the losing end today. On the contrary he said, geology can get a new lease on life with all the things one needs to know today about the whole Earth system. Geology operates with different time scales. For IGBP perhaps it is difficult to sell the last frame of the last 100 million years to the politicians, but if we shift to the last 10 million years there are many new opportunities. Off-shore technologies that have been developed for oil have prompted new advances also in science, for example ocean drilling techniques, which are useful for oceanography; computer technology and satellite imaging have also developed. Technology does not stand still, and this is an important driving factor for science, including the geosciences. Barry Heywood said he tended to agree with this. As for the Dutch case, Shell was instrumental in assisting scientists develop their research program, contributing stored data that gave excellent overviews of what deposits exist in Antarctica. The problem is that scientists are not always good at translating their results into a language that makes sense for broader audiences.

Further Internationalization called for

The question of data banks and the need for wide accessibility of existing information was also discussed. Jim Barnes saw this as a critical point for the future, urging that a pooling of resources and much more far-reaching internationalization is called for. Olav Orheim reminded participants that data was not a matter of straight forward use or non use.

There are different levels of technological development amongst countries, which means that the scientists of third world countries, for example, are at a disadvantage when it comes to "milking" data banks. Here an element of prestige is involved also, so that one has a differentiation within the world's scientific communities. Thus there emerge different cultural perceptions about data, with smaller or poorer countries being at a disadvantage. It is also noteworthy that there is no IGBP data base.

Jim Barnes mentioned how in the US, Senator Nunn has been propagating for a civilian conversion of military computer power, to increase the capacity to handle environmental data. Bruce Davis remarked that, to a political scientist it is clear that economics will not leave Antarctica. To think otherwise would be naive. Moreover, in view of the skew in data availability, favourable to rich countries and unfavourable for developing countries, it is not so strange that Third world nations continue to look with suspicion upon the industrial nations' virtual monopoly regarding computer power and data handling facilities. At the same time, at the other extreme we now have a data-overload; the proliferation of all kinds of data is clogging systems, to the detriment of high level analytical and theoretical work.

Olav Orheim suggested that, in view of current realities a division of labour should by recognized, where the richer nations take on the main responsibility for basic research, which developing nations have no capacity to do as efficiently.

PART V
Panel Discussion and Plenary

CHAPTER 10

Multi-disciplinary and Multi-country Perspectives

The panel was asked to briefly highlight some points of particular significance as reflected by the experience or aspects of activities in which each speaker is currently involved.

* *Riita Mansukoski*, a scientist who is Secretary for the Polar Commission of Finland at the Finnish Ministry of Trade and Industry focused on the three-country model of cooperation that has been developed in Scandinavia, involving Finland - Norway - Sweden. She pointed out how this is a model that small countries can use to facilitate the development of distinct national research programs while pooling resources and entering into a joint venture for logistics purposes. On the other hand logistical cooperation offers scientists an excellent chance to develop cooperation in science, too.

* *Paul-Christian Rieber*, Managing Director of G. C. Rieber and Co. AS, which is the parent company of Rieber Shipping AS, emphasized the commercial dimension of polar activities, calling upon scientists to put pressure on their own country's politicians to use limited funds more rationally and efficiently. This may be done by putting market criteria before nationalist pride. The latter tends to make people less cost conscious. Moreover, nationalism may lead to blind spots when it comes to equipment and safety measures. Rieber also took up the way the Antarctic today is used for many different purposes, not only science, and went on to discuss the implications this multi-purpose use principle has for shipping and other modes of transportation in the polar world.

* *Jan H. Stel*, Director of the Netherlands Marine Research Foundation (SOZ), reviewed his country's experience in developing research in Antarctica. He pointed

out how a high degree of competence in one or more basic research fields may be exploited to pursue a particular niche of considerable importance - in this case marine biology. The Dutch model also involves cooperation with several countries, emphasis on advanced technology, and a recognition of the potential of a European Strategy in Antarctic R&D. The approach differs from the more traditional one of each country mounting its own national expedition in order to qualify for the Antarctic "club".

*   The European dimension was taken further by *Jarl-Ove Strömberg* who is professor of marine biology at the Kristineberg Marine Biological Station, a facility run under the auspices of the Swedish Royal Academy of Sciences, and located on Sweden's West Coast, just above Göteborg. Strömberg is also President of he ICSU-affiliated Scientific Committee on Oceanic Research (SCOR), and played an active role in promoting research cooperation through the European Science Foundation, which has a polar research program EPOS, utilizing the German research vessel "Polarstern". He related his experience as scientific leader of one of the legs of such an expedition a few years ago, and filled us in on the multitude   of organizations, projects and initiatives. Europe is increasingly becoming an actor to be reckoned with in various fields of polar research.

Management and Logistics

Riita Mansukoski's presentation prompted further discussion on the impact of economic constraints and environmental pressures. It was found that the managerial dimension of this problem needs more attention. Economic constraints, environmental concerns and international research cooperation all point in the direction of the need to pool resources and set up joint ventures involving several countries. On the other hand there are various factors that put hinders in the way. Among these are the fact that polar expeditions remain manifestations of political will, and here - at the level of politics - there will be many differences and conflicts to overcome. Secondly, there are differences of commercial practices, and cost-benefit analyses. Thirdly, there exist differences of research traditions and interests which put constraints on what is possible to realize in terms of multilateral cooperation. Finally there is the language barrier, as well as social and cultural differences that put hinders in the way of smooth and effective international cooperation at all levels - planning, logistics, managerial and in the field. For these reasons, it was held that multilateral cooperation works best when it is pursued on a regional basis, involving nations that stand near to each other in terms of culture, language, institutional arrangements and politics. The Scandinavian model was found to bear this out. At the same time it is noteworthy that this cooperation is limited to transportation as a part of logistics and does not carry through to research programs as such.

Logistics has a hardware side and a software side. The latter is the more important, covering the know-how and competence of the people operating vessels, the ship and crew that are there to support the scientists, the helicopter pilots, technicians, consultants, etc. It was pointed out that a ship should always be under the command of the captain, and not of the scientists. Especially in all aspects for matters of safety and in critical situations, the captain must have the last word. Of course the captain and the rest of the crew shall do their utmost to fulfil their requirements of the scientists. It is obvious that social and cultural factors, in many ways, permeate the software side of the logistics equation.

ISBERG, SEDDA DEN 13 NOV. 1902.   *Skiss af förf.*

Iceberg seen 13th Nov. 1902 during the Swedish Antarctic expedition led by Otto Nordenskjöld. Sketch by S. Duse

On the hardware side Paul-Christian Rieber pointed to various types of vessels and vehicles to be considered:

- government icebreakers
- military vessels
- multipurpose research vessels
- bulk carriers and reconstructed supply vessels
- passenger vessels
- helicopters
- planes
- snow-mobiles

In addition there is the question of camping equipment and appropriate types of fuel to be carried.

It appears that the end of the cold war, together with the new environmentalism and the minerals ban in Antarctica, plus budgetary constraints, have altogether converged to increase cost consciousness and made available new solutions to old logistic problems. Thus Russian vessels may now by contracted, and there are dual use research-cum-tourist vessels available. Paul-Christian Rieber insisted that more can be done today with less money if the multipurpose use principle is creatively implemented in logistics. Needed are, not only strong and efficient vessels, but even more important is the know-how of the crew. Vessels must be adapted to the crews and visa versa. Moreover, maneuvering in icy waters is different from one region to another, which means that experience of a particular region is important.

Rieber summed up six points by way of conclusion:

- political missions in Antarctica will continue (cf ice core drilling - operations, where nationalism enters in), and the big nations will prefer to fly their own flag on their vessels.

- increasing cost consciousness will be observed in all countries, with increasingly a total budgeting approach (i.e., weighing in all costs, including repairs, insurance and reconstruction).

- specialization is increasing, both in logistics and in science; at the same time increased flexibility and package solutions.

- more cooperative efforts will develop in future, both with regard to expeditions to one and the same area, as well as between different institutions working on the same subjects, and cooperation between different functions, like science and tourism.

- there will be more long term commitments in Antarctica.

- there will be a greater emphasis on comfort, service and safety during expeditions; safety requirements relate to both people and environment.

The Dutch Approach

The Dutch approach not only reveals the importance of cooperation on the international level, but also on the national level in one's own country. In this case it was a matter of finding the proper mix of national research institutes and university departments. For the first Dutch Antarctic Expedition in 1990/91 a vessel was hired and a part of the Polish research station. At the policy level cooperation with other countries proved useful in bargaining for national funds. Scientists also piggy-backed on the interest in Antarctica created by the international environmental movement and the media.

After a Cabinet decision to allocate HFl 1,8 million per year from 1989, the scientists pursued a number of theoretical lines that had been laid out - glaciological and climatic research, marine science, and oceanography, earth sciences, and studies on environmental impact of scientific operations to minimize the effect of human intrusion. An important point all along has been the emphasis on basic research, where the Netherlands is strong. The fact that the SOZ has a business-minded approach and is able to orchestrate scientific work done in various institutional locations also made it easier for a small group, using a small budget to obtain maximum results. It was a matter of mobilizing different actors in the Netherlands quickly, obtaining support by getting the ear of the politicians and developing a variety of international contacts.

Jan Stel also pin-pointed some future trends that he saw on the horizon:

- fewer research stations, with the closing of "ideological" stations, including the one run by Greenpeace.

- better geographical distribution of research stations needed.

- the influence of the EIA's will be an important factor.

- an increasing number of protected areas, especially in the sphere of marine regions.

As far as logistics is concerned, Stel saw an increasing need for bilateral coordination, international cooperation, and pooling of resources, greater cost effectiveness, and a minimization of environmental impacts. All of these can be counted as performance indicators in future.

The European Arena

The European scene presents a maze of alphabetical letter combinations, also in the field of polar research. Jarl-Ove Strömberg gave an overview.

In 1984 the European Science Foundation (ESF) decided to set up a number of "networks". Seed money for planning was provided to launch the European Polarstern Study (EPOS), using the German Alfred Wegener Institute-based research vessel "Polarstern" from Oct 1988 to March 1989. In all more than 100 scientists from 18 nations participated (two of them South American nations). The net effect was not only scientific data and jointly authored papers but also significant cross-fertilization of ideas and increased cooperation and contacts between European marine science

Major Players in the Organization of International Programmes

ORGANIZATIONS

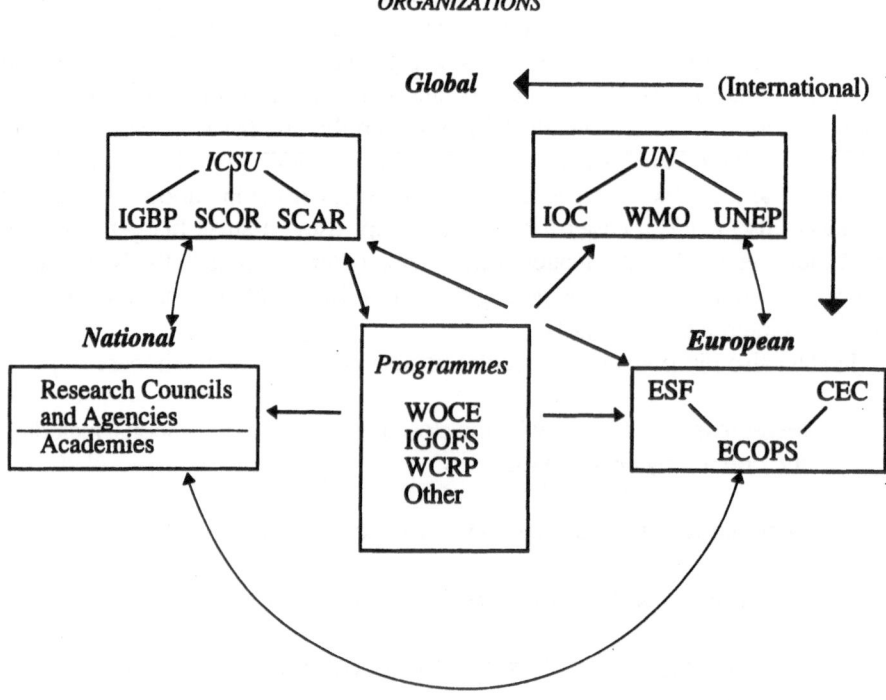

[acronyms are after table of contents]

institutions. Germany carried the costs for the ship, while the ESF funded subsequent workshops and a final symposium in May 1991. The Commission for the European Community (CEC) provided the money for helicopter costs during the cruise.

A second European network has been developed around ECOPS, the European Committee on Ocean and Polar Sciences. This group evaluates and can recommend projects for funding by its parents bodies, the ESF and CEC. ECOPS comprises both "old hands" and new players, among the latter Italy and the Netherlands, for example. It thus constitutes a new forum where an overall European polar research strategy may be discussed, to determine the best use of scientific personnel and instrumental and logostical facilities. Since ESF has limited funds, ECOPS may also be important for raising financial support. The net effect is the growing influence of guide-lines for participating countries, the orchestration of research agendas, and the promotion of cooperation and pooling of resources. Thereby large scale supra-national programs are born. The creation of pan-European vessels has been discussed, but this is not feasible today. Polarstern will continue to play a key role. It may be noted at this point that Professor Gotthilf Hempel of the Alfred Wegener Institute for Marine and Polar Research (AWI) in Bremerhaven is the chairman of ECOPS.

In addition to bilateral cooperation on a regional basis, as in the Scandinavian model, national research councils and academies of science thus have at least three other points of interaction at the international level. These are the ESF, the CEC (and thence ECOPS) and ICSU/SCAR and SCOR. At the European level ECOPS also interacts with the GLOBAL network comprising SCAR and SCOR, SCOPE, both within the ICSU-family, and via the UN and thus the intergovernmental system where we find Unesco with its IOC, WMO, and UNEP. Specific programs are the World Ocean Circulation Experiment (WOCE) run by SCOR and IOC, the Joint Global Ocean Flux Study (JGOFS) run by SCOR, the Land-Ocean Interaction in the Coastal Zone, and others within the IGBP directly under ICSU, and the World Climate Research Programme (WCRP) of WMO with its many sub-projects (for an unpacking of the acronyms see the Glossary p. xi).

With the exceptions of France (Indian Ocean sector) and Italy (Ross Sea) the European nations have very much concentrated their efforts on the Atlantic sector (Weddell Sea) and just west of the Antarctic Peninsula. There is however, a growing interest to investigate the Bellingshausen Sea, so we may see a change in focus in ocean studies on the part some European nations in the next few years.

In sum the panellists pointed to at least five types of international collaboration that will probably develop in the future:

- the Scandinavian three country model

- the European Programmes initiated and/or supported by the European Committee on Ocean and Polar Sciences (of CEC and ESF)

- the ESF network for polar science

- bilateral specially focused ventures with trade offs, like the Dutch-Polish co-operation

- participation in international programmes like IGBP will continue to be important

In Europe the net result may be the development of a large supra-national program that will set guide-lines for participating countries. How loose or firm such a program might be would depend on the individual country commitments and the ability of a lead agency. For scientists in smaller nations, direct communication with a number of influential countries (having resources like ships, research stations, etc.) entice them to cooperate because this is probably the quickest way to get results on intermediate project scales. The European bureaucratic apparatus requires much greater lead time.

Eurocentrism and/or vs. SCAR?

During the discussion it once again became clear that the development of a truly international research station may remain something of a utopian dream. Apart from the political problems involved, there are differences of culture and national style as well as latent and in certain cases overt nationalism that plays in. Especially the British, who have a long tradition of research in Antarctic, with a newly revitalized organizational structure with large outlays for new infrastructural arrangements to facilitate greater cost-effectiveness, remain sceptical. In the Netherlands, Finland and Sweden the expectation is that in the future new technologies will drive forward further integration. It is possible that this reflects managerial views but the current discussion concerning a united Europe of the future, and the evolution of a corresponding supra-national S&T policy framework is not without interest in this context. Bruce Davis was quick to point out the essentially Eurocentric nature of this vision of the future, noting that it leaves out many nations, especially those in the Third world. Jim Barnes intimated that nation states will continue to be the main actors, even in the future. This is also the case, ultimately, in such constellations as CERN, the European molecular biological laboratory in Heidelberg and ESO, the European Southern Observatory with its modern facility in Chile and headquarters in Munich. Still, there will be a reduction in intensity of the competition fanned by the East-West tensions of the Cold War era and its aftermath, which was responsible for a lot of the early station building activities in Antarctic. The Dutch approach and its success in opening a new road to full membership in the ATS, i.e., without building a research station, is a breakthrough. It carries a precedential aspect that will permit further nations to participate under new conditions. At the same time it would be a pity if this is limited to European initiatives. Also, the relationship between SCAR and ECOPS is unclear. It is important that SCAR's role be strengthened in parallel with the strengthening of Europe as an actor. Failing this, SCAR may find itself increasingly overshadowed, by-passed and certainly at odds with the new supra-national structures that have developed. SCAR, after all, not only stands for quality, but also for a more encompassing internationalism.

Technology as a Driving Factor

Nigel Bonner saw the current trend of new forms of cooperation as a temporary one. Already there is a countervailing tendency of fragmentation of efforts, driven by particular interests. Furthermore, if one looks beyond the science, nationalism of countries continues in other exploits like skiing expeditions to the South Pole, dog sledging to the same destination, and private yachting in Antarctic waters. All these lead to further environmental impact and safety problems which lie outside the control of scientific research station managers. They are a significant factor to be reckoned with, especially on the Peninsula.

Olav Orheim noted how new technologies and changing trends in logistics makes it more attractive for senior scientists to participate in expeditions and field trips. Here both the UK and the French airstrips are important, facilitating a lot more good science per investment of time and money. Bruce Davis commented on this to note that the French Dumont d'Urville airstrip is potentially of interest to the Australians too. Earlier the Australians have had their own debate about a similar airlink from Hobart directly to one of their own bases. Some businessmen showed an interest in such a venture, but it has never gotten beyond speculation.

Ingemar Bohlin took up the point about how governments spend money in ways that is not cost effective from the point of view of science. In his estimation this will continue as long as science can be used as a vehicle for other goals, be they political or otherwise. Governments take an instrumental view of science. If science can function as symbolic capital in a number of other contexts, there may be a fruitful tradeoff between science and politics. The cold war tensions spurred this equation. The question now is whether the politics of environmentalism is sufficient to supplant it.

Barry Heywood agreed that the collapse of the East-West conflict has affected a number of programs, especially that of the former USSR, where economic crisis now also has led to further constriction of Antarctic science. However the collapse of the East-West conflict will not affect the UK, France and the Scandinavian countries. Here the cost-effectiveness of new technologies remains an important driving factor; and so does the basic research interest.

Externalism and Internalism

The plenary session ended after a further discussion concerning conservation measures relating to seal stocks and krill. It was noted that SCAR has played a progressive role here, pointing to the need of zoning and management zones at a very early stage of the history of the ATS. This was not possible to implement, and instead one had the designation of Specially Protected Areas (SPAs) and multi-use areas. Some of this has to be further fleshed out within the Environmental Protocol. A scientific workshop in Cambridge in July 1992 will probably take up some of the difficulties pointed to during the symposium in Göteborg.

By and large externalist aspects of the dynamics of Antarctic science dominated during the two days of discussion. This will have to be complemented by further research on internalist factors, such as disciplinary differentiation, specialty formation and the changing pattern of integration with science more broadly on a number of international research fronts.

PART VI
Four Symposium Papers
and a Review of SCAR

CHAPTER 11

The Science/Politics Interface in Development

*Nigel Bonner**

Perhaps more than anywhere else in the world, science has been inextricably linked with exploration and development in Antarctica. Many of the earliest voyages of discovery were government-sponsored, though often primarily undertaken with hopes of commercial reward. Scientific research was a normal part of their programmes. When Captain Cook set off on his voyages of discovery that were to climax in his first crossing of the Antarctic Circle in 1773, he carried with him the famous German naturalists Johan and Georg Forster who made and described some fascinating collections of scientific curiosities. Charles Wilkes, leading the United States Exploring Expedition in 1838–42, took with him no fewer than seven scientists (and two artists) to conduct a broad scientific programme, though only one of these actually accompanied the Antarctic portion of the cruise. Commercial sealing voyages, such as those of James Weddell between 1819 and 1827, also made valuable scientific contributions.

Purely private, non-commercial expeditions were few, but made a disproportionately large scientific contribution. Outstanding amongst these was Otto Nordenskjöld's Swedish South Polar Expedition of 1901–04. Despite the loss of his ship, the *Antarctic,* Nordenskjöld's privately funded expedition was a great success. Nordenskjöld himself conducted the geological investigations, Carl Skottsberg was the botanist, K Anderson the zoologist, Erick Ekelöf the bacteriologist, while S A Duse and G Bodman carried out mapping and hydrographical work.

Of the other great Antarctic explorers of this period, Amundsen, Scott, Shackleton and Mawson, only Amundsen's expedition failed to yield important scientific results.

The period between the World Wars saw politics beginning to play a major role in the Antarctic. Great Britain had made a claim to an area known as the Falkland Islands Dependencies in 1917. This was followed by claims by New Zealand in 1923, France in 1924, Argentina in 1925, Australia in 1931 and Norway in 1939. These claims were not universally recognised - indeed, this was not possible, as two of them related to virtually the same geographical area - and inevitably political tensions developed. These were most acute in the Antarctic Peninsula region where the claims of the United Kingdom and Argentina (and after 1950, Chile also) overlapped.

In 1943, at a time when activities of any sort in the Antarctic were at a minimum, the United Kingdom sent a naval party under the code name "Operation Tabarin" to occupy bases in the South Shetlands and the Antarctic Peninsula. This was to counter the potential threat that German armed merchant cruisers might use these harbours from which to prey on Allied shipping. When the war ended, Operation Tabarin was reconstituted as the Falkland Islands Dependencies Survey, with a full scientific programme, conducted from a further seventeen research stations set up in the Antarctic.

In 1947 Argentina and Chile both operated meteorological stations in the Antarctic Peninsula. Argentina, it may be noted, had been operating a meteorological station at Laurie Island in the South Orkneys since 1904, which it had inherited from the Scottish National Antarctic Expedition. Political tensions resulting from the conflicting territorial claims in the Peninsula by these three nations mounted steadily, and culminated in February 1952 when an Argentine party opened fire with machine guns on a British party at Hope Bay.

Meanwhile, despite these disturbing political events, science proceeded. The discipline that was to draw together all the nations engaged in Antarctic research, was geophysics. Many geophysical phenomena are best studied at high latitudes, and this had led to cooperative "Polar Years" in 1882-83 and 1932-33. In 1950 discussions took place between influential geophysicists which resulted in 1957-58 being chosen as the third Polar Year as it was to be a period of maximum sunspot activity. This was formalised under the auspices of the International Council of Scientific Unions (ICSU) as the International Geophysical Year (IGY).

Despite the political problems existing between Argentina, Chile and the United Kingdom, and the Cold-War frostiness between the USA and the USSR, the scientists of the twelve countries involved succeeded in working out a programme that avoided duplication and used their resources to the best advantage.

During the IGY there were 47 research stations working in the Antarctic south of Latitude 60°S and a further eight stations on islands to the north of this. To use this network to the best advantage, the scientific programme was widened far beyond geophysics and in 1957 ICSU established the Special Committee on Antarctic Research, later re-named the Scientific Committee on Antarctic Research, SCAR. This had the task of bringing together scientists of the different disciplines and coordinating programmes and exchanging information.

The IGY saw a truce to the territorial bickering in the Peninsula and the development of a remarkable degree of scientific cooperation between the countries involved. International politicians were quick to see that the dialogues being built up, particularly those between the USA and the USSR, could be used for other purposes. In 1958 the USA took the initiative to urge representatives of the twelve nations concerned - the seven claimant states, Argentina, Australia, Chile, France, New Zealand, Norway and the United Kingdom, together with Belgium, Japan, South Africa, the USA and the USSR - to hold a conference to "seek an effective means of keeping Antarctica open to all nations to conduct scientific or peaceful activities there". From this conference was born the Antarctic Treaty, signed in Washington in 1959 and which came into force in 1961.

The Treaty agreed to freeze the situation with regard to the recognition, or non-recognition, of territorial claims; it provided the basis for the development of scientific cooperation without political interference; and it guaranteed the freedom for scientific research throughout Antarctica. The Treaty had little to say about conservation and no mention at all was made of environmental protection, but these subjects were destined to be the main field for the interaction between science and politics.

At the Third Antarctic Treaty Meeting in Brussels in 1964, a set of measures was introduced and adopted which were the first steps within the Antarctic Treaty to provide protection for the environment. These were the Agreed Measures for the Protection of Antarctic Fauna and Flora. The Agreed Measures prohibited the killing of any native mammal or bird; called upon governments to minimise harmful interference with the normal living conditions of wildlife and alleviate coastal pollution; made provision for Specially Protected Areas; and regulated the introduction of non-indigenous species.

The Agreed Measures were based on a set of conservation principles prepared by SCAR and it is significant that in an early recommendation (I-IV) the Treaty had urged the continuation of SCAR's advisory work that had "so effectively facilitated international cooperation in science". Recalling that SCAR is a non-governmental organisation, it was even more remarkable that the Treaty, in pursuing a policy of environmental protection, repeatedly called on SCAR to provide it with advice. In Recommendation VII-2, for example, SCAR was invited to review existing and proposed Specially Protected Areas.

As the protection of the environment was increasingly seen as more important than specific measures to protect fauna and flora, the requests for advice from SCAR grew more frequent. Recommendation XII-3 asked for SCAR's advice on the categories of scientific and logistic activity that might have a significant effect on the Antarctic environment, and asked SCAR to elaborate relevant assessment procedures. This first step towards environmental impact assessment marked a significant new departure for the Treaty and SCAR's response was not universally welcomed by all parties, though four years later, at the XIV ATCM, a Recommendation on EIA was adopted.

Several of the requests for advice were couched in diplomatic, rather than more rigorous scientific, language, and thus presented SCAR with a challenge in responding to them. SCAR's method of response was generally to set up an *ad hoc* group, or a group of specialists, who at one or a series of meetings would try to develop a paper containing the advice. The Treaty meetings were, in general, two-yearly (in odd-numbered years), while SCAR met in even-numbered years. This provided a time slot of about a year between the Treaty request for advice, and SCAR's next meeting. In general this was too short a period for the setting up of the group, holding its deliberations and preparing a report that could be examined by a full meeting of SCAR. A consequence of this was that often advice formulated by a small group was examined only by the SCAR Executive (President, two Vice Presidents, Past President, Treasurer and Executive Secretary) before being passed to the Treaty. In practice, this did not seem to cause much difficulty.

The route of advice between SCAR and the Treaty was complicated since SCAR, being a non-governmental organisation, had no formal contact between its Executive and the Treaty, which in any case lacked (and still lacks) a secretariat. Formal contacts could only be made by SCAR National Committees with their appropriate governments. Very often matters at the Treaty relating to the environment were discussed in advance by a small group composed of New Zealand, Norway, the UK and the USA. Frequently, SCAR advice was passed from the SCAR office, which is situated in the UK, to the head of the UK delegation at the Treaty, but this was not invariable and any party had the opportunity to forward SCAR papers.

The 1980s saw a great increase in environmental awareness in the developed world. The growth of the "Green" movement did not neglect Antarctica. Greenpeace, having successfully concluded its campaign against the killing of harp seals in the north-west Atlantic, turned its attention south and in the 1986-87 Antarctic summer set up a four-man station at Cape Evans on Ross Island.

Meanwhile, the Antarctic Treaty parties had been concerning themselves with the question of possible minerals activities in the Antarctic. When the Treaty was being negotiated in 1959 the question of minerals was discussed but no reference was made to it as it was felt at that time it would be premature. The quadrupling of the price of crude oil by OPEC in 1973-74 aroused interest in possible Antarctic minerals and the Treaty asked SCAR to "make an assessment on the basis of the available information of the possible impact on the environment of the Treaty Area and other ecosystems dependent on the Antarctic environment if mineral exploration and/or exploitation were to occur there". SCAR submitted a report to the IX ATCM (the "EAMREA Report") but part of this was not politically acceptable (mainly to the USSR) and the Treaty then set up its own Intergovernmental Group of Experts, which produced a parallel report.

At the XI ATCM in Buenos Aires in 1981 the Treaty agreed to hold a Special Consultative Meeting to discuss control of mineral resources development. SCAR was again referred to for advice, and again responded in a report "Antarctic Environmental Implications of Possible Mineral Exploration and Exploitation (AEIMEE)" in 1986. Two years later, in Wellington in 1988, the Convention on the Regulation of Antarctic Mineral Resource Activities (CRAMRA) was adopted by consensus.

I shall not go into any details of CRAMRA, other than to say it made provision for elaborate environmental safeguards should minerals activities ever take place.

Despite this, it was regarded as anathema by the Green movement. Environmentalist groups, coordinated by ASOC, the Antarctica and Southern Ocean Coalition, set out to destroy CRAMRA and designate Antarctica a "World Park". Strong public support for ASOC's aims, particularly in Australia and France, led to these two countries announcing that they would not sign the Convention. Since the signatures of all the seven claimant states were needed for CRAMRA to come into force, this meant that CRAMRA was moribund, if not dead.

The new environmental thinking that had gone into CRAMRA had alerted the Treaty to the realisation that its own environmental measures were in need of revision and rationalisation. At the XV ATCM in Paris in 1989 it was agreed to hold a Special Consultative Meeting to discuss proposals relating to the comprehensive protection of the Antarctic environment. The Parties met in Chile in November 1990 and in Madrid in April and June 1991. Initially there was something of a confrontation between a group of four, Australia and France, joined by Belgium and Italy, who wanted a new Convention dealing with Antarctic environmental protection, and the majority, led by the USA and UK, who were in favour of a protocol to the existing Treaty. New Zealand held a somewhat intermediate position.

By the end of the first Madrid meeting, it had been agreed that the revision of environmental measures under the Treaty should take the form of a Protocol. Unfortunately, some rather undiplomatic manoeuvering by the United States meant that the Final Act was not signed at Madrid, though it is confidently expected that this will be done before the next ATCM in Bonn in October. [Note: It was adopted and signed 4th October 1991]

The Protocol embodies all the existing environmental protection measures through a series of Annexes, which can be readily updated or revised. In its first article, it designates Antarctica as a natural reserve, devoted to peace and science. Operationally, the Protocol would function through a Committee for Environmental Protection. This would be an advisory body, reporting to the full consultative meetings. In carrying out its functions the Committee has to consult with SCAR, CCAMLR (the marine resources convention) and "other relevant technical environmental and scientific organisations". SCAR, and other organisations, may be invited to participate as observers at meetings of the Committee. The report of each of the Committee's meetings "shall cover all matters considered at the meeting and shall reflect the views expressed". (It is not entirely clear whether SCAR, as an observer, has the right to have its views reported.)

The special status of SCAR in providing advice to the Treaty is thus preserved in the Protocol, but there has been some watering down of wording put forward by the Swedish delegate at the Chile meeting. Originally this stated: "In carrying out its functions, the Committee shall have regard to the work of the Scientific Committee on Antarctic Research .... To that end, SCAR shall be invited to present their views and to comment on proposals within their competence put forward by the Committee. Such comments shall be presented together with the report from the Committee." The wording finally agreed is, as noted, less supportive of SCAR.

A legal mechanism for protecting the environment will work only if it is based on sound scientific advice. Where the Antarctic is concerned, SCAR is the organisation that can provide the best access to such advice. SCAR uniquely embodies the individuals with practical experience of Antarctic conditions and of carrying out scientific research programmes under these conditions.

These research programmes represent not only the academic interests of the scientists concerned. They are of general importance to the world. No one now doubts the importance of research in the Antarctic in drawing attention to the ozone hole, for example. Similarly, research on Antarctic ice cores has provided us with a series of datable samples of the Earth's atmosphere, stretching back for millennia, and throwing light on the role of carbon dioxide in global warming. On a shorter time scale, Antarctic snow chemists can demonstrate global background levels of various substances, ranging from the lead derived from industrialisation to the radioactive fall-out from atomic explosions. The role of the Antarctic in driving the world's weather systems has been revealed by meteorological and oceanographical research.

I could continue in this vein for some time, but putting it more simply, the Antarctic constitutes 10% of the world's surface, and we cannot afford to ignore this important part of our planet.

This is the justification for scientific research in the Antarctic, much of which relates to our understanding of our environment. At the same time, no one doubts that the Antarctic environment needs protecting, and this is the function of the various measures, culminating in the Protocol for Environmental Protection, adopted by the Antarctic Treaty.

Is it paranoid for an Antarctic scientist to have some doubts about how environmental protection may affect research? I do not think it is. Already some regulations for environmental protection seem based more on what is appropriate elsewhere in the world than what is needed in the Antarctic.

I will give an example based not on Antarctic Treaty legislation, but on the International Convention for the Prevention of Pollution from Ships, 1973 and its Protocol of 1978 (MARPOL 73/78). At its meeting last November, the Intergovernmental Maritime Organisation, IMO, designated the Antarctic Ocean a Special Area under MARPOL. This designation, which imposes special restrictions on the discharge of oil, garbage and sewage, was designed to control pollution in enclosed sea areas where there was a great deal of shipping. Existing examples are the Baltic Sea, the Mediterranean and the Persian Gulf. This designation was wholly unnecessary in the Antarctic, which constitutes the most open and energetic sea in the world, and where the density of shipping is the lowest of all the world oceans. The provision of reception facilities for oil, garbage and sewage is not only impracticable, but is also undesirable, in the Antarctic.

Will this designation as a Special Area under MARPOL affect Antarctic research? Possibly not. Ships are regularly being brought up to MARPOL standards and certainly no oil should be released into the ocean anywhere in the world. But ships may find themselves in breach of the regulations inadvertently. One of the Special Area regulations prohibits the discharge of sewage within 12 nautical miles of land. But what happens when a supply vessel unloading at a base is constrained to remain alongside because of ice conditions? Few ships can store sewage for weeks on end, and there will be no reception facilities for the sewage ashore. To add a further twist to the situation, the shore base, under existing and proposed regulations, will be able to discharge its sewage to the sea, while the vessel under the MARPOL rules, and rules to be adopted by the Treaty and its Protocol will be forbidden to do so.

It is no help in the Antarctic, as elsewhere, for the law to be seen to be an ass. Proper concern for scientific, and practical, advice at the drafting stage could have avoided this situation. The advice was available, but was dismissed for political reasons.

A further question arises as to how the advisory process will operate when, as is expected, the Protocol comes into effect and the Committee for Environmental Protection examines matters of environmental concern. The Committee will consist of delegates from each Treaty Party. These, presumably, will be diplomats. They will be supported by scientific advisors recruited either from the appropriate government departments or independent institutions.

Because of the limited number of persons with experience in the Antarctic, it is almost inevitable that these scientific advisors will be people who, at another time, might be serving on a SCAR group, providing advice to the Treaty. Such a duplication of the source of advice might seem to be no bad thing - scientists should give the same advice whatever hats they happen to be wearing - but I believe that this may not always be the case. Scientific advisors within the Committee must speak through their head of delegation - a government representative - and unpalatable advice will inevitably be filtered out. SCAR, of course, is not free of political pressures, but these are in general subdued as they were in the run up to the IGY. The advice of a multi-national group set up by SCAR is, *a priori*, likely to be more disinterested than that from an advisor within a national delegation with a national policy to put forward.

There is a danger that the Treaty might suppose that one advisory body - the Committee for Environmental Protection - is enough, and cease to ask for advice from SCAR. For the reasons I have just described, I believe this to be inimical to the best interests of science and the environment in the Antarctic.

There is another, and perhaps more immediate danger. Even if the Treaty continues to ask SCAR for advice, there is a possibility that SCAR may be unable to respond. SCAR's main role, as I noted before, is the coordination of science and the exchange of information. Environmental matters are peripheral to its main theme. Providing a response to the Treaty almost always requires the setting up of a workshop - experience has shown that trying to provide and collate a reasoned response by correspondence is futile. The costs of a meeting of a multinational group of, say, six persons meeting for five days, even when they provide their services free, is likely to be around $10,000. This is the same as the total annual subscription of a SCAR member and represents a significant drain on SCAR resources. SCAR currently has been unable to provide properly considered advice on questions relating to protected areas and environmental monitoring because it has not been possible to find the funding for meetings on these subjects. A request for funding for a meeting to consider data storage and analysis (in response to a request in Recommendation XV-5) would by itself have absorbed all of SCAR's meetings funds for one year. It is clear that if SCAR is to continue to act as a source of advice to the Treaty, some arrangement must be made to meet the costs of this. If, as is likely, the Treaty equips itself with a Secretariat, this problem may be more easily capable of solution.

I believe that the provision of independent scientific advice to the Treaty from adequately experienced sources is of prime importance not only to the protection of the Antarctic environment but to the proper care of the world environment as a whole. International politicians must consider carefully whether they will base their policies on vociferous populist groups whose experience is mainly in the field of manipulating the media, or on the band of scientists whose experience of the Antarctic has included long sustained efforts at environmental protection. As a member of this latter group I have no doubt which is the most prudent course.

CHAPTER 12

Science in an Environmentally Regulated Continent

James N. Barnes*

Introduction and Summary

One could answer the question posed by this topic very briefly by saying, science will be in first place, as it has been for a long time in the Antarctic.

Among the primary reasons why the environmental community has worked so hard to refocus governments on protection as opposed to exploitation of the region is that it seemed imperative to preserve the qualities that make Antarctica a global laboratory of great importance.

Thus, I would argue that the proper perspective provided by the conclusion of an Environmental Protection Protocol to the Antarctic Treaty is that science will be even more important than before, and that its possibilities are greater. I doubt that the Protocol will block any significant science from being carried out. Rather, the practical process of implementing the Protocol will help governments and scientific organizations focus more clearly on priorities for their scientific programs. The results will be (1) more efficient science, (2) more effective science, (3) more money for science, (4) more long-term monitoring programs, and (5) more directed research, for example, of the sort needed to effectively implement the "ecosystem as a whole" principle that lies at the heart of CCAMLR.

Among the likely positive benefits of putting aside the minerals question is that the political will to support globally significant science and long-term monitoring programs in a cooperative way will increase. This should result in greater sharing of bases, facilities and logistics, and a consequent decrease in environmental inputs caused by a redundancy of these support facilities.

If more globally significant science were the result, and not only as a theoretical phrase used at festive events, but as a practical reality, this would be a very good thing. The increased "transparency" of decision making will also make more clear the responsibility of scientists, for it is a truism that science is value-free, but scientists are not. As Lloyd Timberlake has written: "Science has a lot to answer for, it has invented weapons that can destroy Nature as we know it; it has produced chemicals which pollute our water system and our atmosphere. It is, in the democracies, up to the electorate and those they elect to harness and direct science to make that interplay between nature and ourselves more supportive of both nature and ourselves."

Inevitably, there will be many practical aspects to the implementation of the Protocol that managers of Antarctic programs will have to take into account. In turn, these practical aspects will change to some extent how science is done — in particular, how it is serviced. The Treaty parties have been moving in this direction for several years. The Protocol has not been drafted in a vacuum; it ratifies that trend, and casts it in legally binding language, which is a major step forward. But the drafters and negotiators put in reasonable time frames for implementation, which will limit the burden on ongoing scientific programs.

I know that there are fears in the scientific community that they will be "over-regulated" as a result of the Protocol. Richard Laws wrote a piece on this just recently, for example. But this is a misplaced fear. I doubt that there is any legitimate science that will be impeded by the conduct of an environmental impact assessment.

It is true that research requiring massive logistical support, especially if it had the potential of opening an area to other types of activities, will be scrutinized very closely, and properly so. For example, core drilling in the Dufek Massif would not, in itself, cause a problem, but a hardened runway to service the research might be. There also may be beneficial economic implications that result from the need to conduct an environmental impact assessment. A government might reconsider whether a research proposal of questionable merit should be given priority use of scarce funds. For example, if blow-out preventers were required on drilling operations in the ocean, to protect against the chance of a hole being bored on structure, a manager might conclude that it would be too expensive. Only time will provide the answers to these scenarios.

Role of Independent Scientists

I want to say a few words at this point about the role of SCAR and independent scientists in general. SCAR has had the responsibility of initiating, promoting and coordinating Antarctic science. It is the one international, interdisciplinary, non-governmental organization that is able to draw on the experience and expertise of scientists across the boundaries of nations and subject matter. SCAR's advice to the Treaty parties during the last thirty years has been excellent. Although environmental groups have sometimes criticized some SCAR reports, and its failure to weigh in on some contentious subjects, such as the construction of a hardened air field at Point Geologie, in general its advice on protecting the ecology and environment of the region has been ahead of government thinking at the political level. Several years ago, for example, SCAR made some good proposals on the question of environmental impact assessments, which took governments several years to agree to, and not in so strong a form as SCAR had proposed.

In working out the practical implementation of the Protocol, it is imperative to maintain a strong position for the provision of independent scientific views, through SCAR as well as other pertinent sources of such advice. The SCAR Group of Specialists on Environmental Affairs and Conservation was formed in order to provide such advice. In my view, SCAR should not, as a result of the Protocol, be replaced as the main body for advising the Antarctic Treaty System on the scientific aspects of environmental protection. NGOs look forward to working closely with SCAR in the future.

Without attempting to provide an exhaustive list of other sources of independent advice, there are obviously a range of other scientific groups, institutes, as well as individual scientists, with expertise to be utilized in the Antarctic. In a sister body, there is the CCAMLR Scientific Committee, and several other scientific groups, all looking into the implementation of CCAMLR's innovative "ecosystem as a whole" principle. Their expertise and present data bases should be extremely useful in the implementation of the new Protocol. Also, the international NGO environmental community includes a number of excellent scientists, who are available to the Antarctic Treaty System.

Globally Significant Science

I will speak briefly about the unique opportunities afforded by the Antarctic for research that contributes to an understanding of problems outside its boundaries. At present, Antarctica is a global scientific laboratory of immense value. The near pristine nature of the region provides a baseline against which we can measure pollution in more populated regions.

The information locked in its ice cap is helping us to better understand the climatological history of the planet.

Antarctica plays a key role in the energy balance of the globe; Antarctic research is crucial in the understanding of global change phenomena, ozone depletion and the greenhouse effect. The International Geosphere-Biosphere Programme (IGBP) has been established under the auspices of the International Council of Scientific Unions. IGBP describes the importance of the polar regions in this way:

The polar regions are very sensitive to changes in the global environment and may act as "warning signals" to changes in the total energy flux into our Earth and to changes in the atmosphere. The polar regions also act as global historians, maintaining records of past global environmental conditions within their permanent ice fields.

The IGBP is coordinating international research efforts in the polar regions, focusing on prime indicators of global change: ozone concentrations, ice cores, polar ice levels, and polar temperatures. It is important that such phenomena be studied without interference from local sources of pollution. The IGBP has identified numerous priorities for its Antarctic components, which may be summarized as:

- detection and prediction of global change;
- study of critical processes that link the Antarctic to the global climate system;
- provision of information on the history of environmental change; and
- assessment of ecological processes and effects.

These investigations will allow humans to understand the interactive physical, chemical and biological processes that regulate the Earth's life support systems, the changes that are occurring, and how those changes are influenced by human activities.

The potential of the Antarctic for extremely important research on global problems is not being fully realized. Human ability to manage and control increasingly varied and intense human activities and their impacts on natural systems is dependent on a much better understanding of the interactions of fundamental components of the global ecosystem. Depletion of the ozone layer, global marine pollution, long-range transfer of pollutants, and climate change phenomena require coordinated studies on a global, interdisciplinary, multi-institutional basis. We all know this.

Antarctica has been a good proving ground for innovative scientific endeavor, such as the ozone depletion experiments and sophisticated analytical work done at high altitudes over Antarctica in recent years. It was as a result of measurements taken by British Antarctic Survey scientists at Halley Bay Station that we learned in 1985 that the ozone layer over Antarctica had been decreasing systematically during the period 1975 to 1985. In 1986 the U.S. National Science Foundation, National Oceanic and Atmospheric Administration, NASA and the Chemical Manufacturers Association organized an Antarctic Ozone Expedition Team, which took measurements on the ground and in balloons. That research led to a number of important papers on the problem and dramatically raised public awareness about the need for action to address the causes of ozone depletion.

Since then, increasingly sophisticated field instruments, research techniques, computers and satellite capabilities, have been used in the Antarctic. In general, these activities are beyond the resources of most nations and any individual research institution. Only through multi-national efforts can we begin to fully realize the benefits of increased understanding of the earth and its systems. Careful long-term planning and allocation of sufficient financial and intellectual resources are required, as well as continuity of support for these large global research and monitoring programs.

In the view of environmental organizations—and I believe also in the view of most scientists — not nearly enough financial support has been committed by governments to the IGBP Antarctic work plan.

It is also worth noting that in the Arctic, there is a new Arctic Science Committee (IASC), with which there should be close cooperation with Antarctic scientists

working on global change phenomena.

Other Scientific Realms

Turning to some other areas of science, the Protocol will help give countries the incentive to cooperate on biologically important questions — questions that are unique to the Antarctic.

Many simple terrestrial and aquatic ecosystems found in Antarctica are available as models that can be used to follow critical biological processes both at the population and community levels. The Southern Ocean surrounding Antarctica plays a key role in regulating global tides, ocean currents, and sea levels. By protecting these waters, the Protocol and its Annexes, particularly the Waste Disposal and Marine Pollution Annexes, will enable countries to implement long-term programs that will help elucidate the interaction of the Southern Ocean and the oceans and climate of the world.

Because the waters of the Southern Ocean are among the most biologically productive in the world, they support a unique, highly adapted and specialized ecosystem. Antarctica is the world's largest wildlife sanctuary, home to over 100 million birds, including seven species of penguins and six species of seals. It is the summer feeding ground for fifteen species of whales.

This huge marine ecosystem is unusual, as one krill species, *Euphausia superba*, helps to support all of the higher species. Antarctic krill is the major food source for five species of whales, three species of seals, twenty species of fish, three species of squid, and numerous penguin and other bird species.

Annual estimates of krill production range up to 15 million tons. *There is an active krill fishery in the Antarctic, which could interfere with the ability of these krill predators to obtain a sufficient diet to sustain themselves and their offspring if it is not properly regulated.*

The Convention on the Conservation of Antarctic Marine Living Resources was negotiated because several Treaty nations were concerned that over-exploitation of krill would lead to the general demise of the Antarctic marine ecosystem. This concern remains. *Due to the lack of information on the numerical and functional relationships between krill and other components of the Southern Ocean ecosystem, it has not been possible thus far to obtain agreement on a precautionary krill cap.* Among other things, there is no agreement on

* the number, size and productivity of krill populations in the Southern Ocean
* the biomass of krill necessary to sustain krill predators and ensure krill recruitment, or
* the biomass that can be harvested safely.

It is reasonable to assume that the total population of krill could sustain the fishery

at its current level, if the fishery were dispersed throughout the Southern Ocean. But the fishery is concentrated in those areas where krill swarms are known to occur. These are also the areas where krill predator populations seem to be the largest. *It is conceivable that the localized effect of fishing in these areas could cause significant impacts on one or more of the dependent populations.*

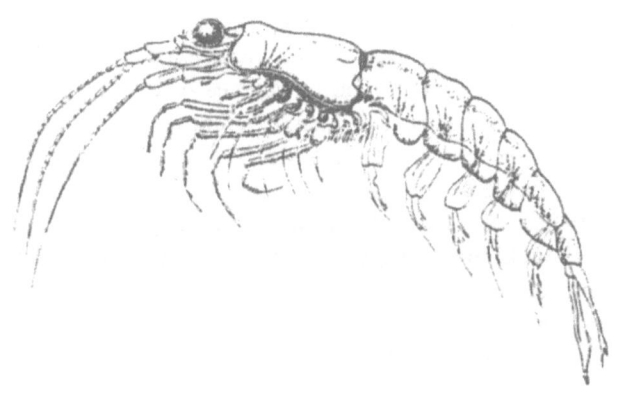

In view of the potential for a rapidly escalating krill fishery, there is a need to gather more data and to analyze critically the data collected so far on krill and krill predators in the Southern Ocean. The new era of cooperation encouraged by the Protocol could be a prime motivator in governments jointly undertaking the directed research programs that are essential in order to provide answers to critical questions such as those listed above, as well as the following:

(a) How do the various krill predators locate and prey upon krill?

(b) Will different harvesting practices, as well as different quantities of harvest, affect predators differently — e.g., is krill availability to various predators dependent only upon total krill biomass, or on variables such as the number, size and density of krill swarms?

(c) How long might it take for harvest-induced changes in krill availability to be reflected in and detected by ongoing programs that monitor selected krill predators? How could/should fishery development be structured to take ac-

count of such time lags?

(d) Is there any reason to believe that current harvest levels or practices may be having adverse impacts on krill stocks locally (e.g., in the South Georgia area), regionally (e.g., in statistical area 43), or throughout the Southern Ocean?

(e) Given available data, what manner and level of krill harvests could be allowed in the various statistical areas with high (e.g., 95%) confidence that they would not have adverse effects on target, dependent, or associated species as defined in CCAMLR?

Given this overall context, one can examine the options coming before the CCAMLR this year, regarding a possible "krill cap" and related precautionary regulations. I believe that it is desirable to have an agreed cap on the krill harvest—even if it is based on unreliable and insufficient data. But the level that is accepted must not be too high and must be acknowledged as preliminary, subject to further refinement. There is a need to keep moving toward a position of information that is sufficient for sophisticated — and correct — judgments to be made about the truly appropriate cap, by region, with whatever restrictions are needed in terms of timing of fishing to protect the breeding cycles of predators. ASOC will introduce an Information Paper on this subject at the CCAMLR meeting.

Conclusion

In conclusion, I submit that the new Environmental Protection Protocol will advance the missions of those national agencies around the world that are conducting research in the Antarctic. The Protocol will neither interfere with nor harm scientific endeavors, but rather will help protect the region in perpetuity so that its scientific potential can be fully realized.

The Protocol has moved the world one step closer to realizing the values of the Antarctic. It is a zone of peace, providing experience in international cooperation. There is now added impetus for the nations of the world to work together, with NGOs in the scientific and environmental communities. When I say "nations" of the world, I mean to encompass *all* of the nations, not just those that have membership in the Antarctic club. In this context, there are some exciting possibilities of international bases being established.

The Antarctic is the world's-only truly demilitarized zone. With the threat of mineral resource exploitation removed, this status will be preserved. It is the best place on earth to monitor and learn about global weather systems, global warming and ozone depletion. We have just begun to appreciate Antarctica's complex environment, and its major contributions to the Earth's life support systems. The Protocol has provided us with a large window of opportunity to pursue this multi-faceted investigation.

CHAPTER 13

The Australian Antarctic Research Program in Focus

Bruce Davis*

Introduction

The principal aim of this paper is to illustrate the manner in which increased political and community interest in Antarctica is shifting the focus of Australian Antarctic research towards environmental management, creating some tensions amongst bureaucrats and scientists as to program priorities and funding allocations (Davis 1990).

The process is ongoing and it is too early to predict an outcome, nonetheless widened public debate about research objectives and allocative efficiency in Antarctic programs, including those relating to the Southern Oceans and Sub-Antarctic island dependencies, has probably resulted in some beneficial effects, but needs to be viewed against the broader context of efforts to restructure Australian R&D effort more generally, which have occurred in recent years (DPIE 1985, Prime Minister's Science Statement May 1989, ASTEC 1990 and 1991). The stated intent is to recruit more personnel into science programs, rationalise and render more effective government R and D effort and increase R&D investment by the private sector, as well as promote the notion of excellence in science (Slatyer 1991). To what extent these efforts will impinge upon Antarctic science specifically is not clear, however the initial signs are hopeful as both the level of funding and program diversity are gradually being expanded (Antarctic Division 1991).

There remain some paradoxes about the manner in which Antarctic science is conducted by Treaty nations. Scientists are prone to argue that the most valuable breakthroughs are likely to occur when committed researchers are left free to examine phenomena of their own choice (Heap 1983, Roots 1986, Quilty 1990). But this ignores the reality that most science is now government funded, where the public interest necessitates close attention to relevance and the distribution of benefits and

* Author presentation on p 71

scientific infrastructure support is directly related to the politics of science (Lowe 1991). Science is *not* value-free, hence what investigators choose to examine and how they analyse and report it are very much bound up with cultural and educational factors, pride and ambition and the power that knowledge professionals come to exert in technological societies (Maynaud 1968, Bahm 1971, Beneviste 1973).

Antarctic science also reflects national aspirations, so it is not surprising that despite the rhetoric of international cooperation, there is much needless duplication of station siting and data collection, especially in the Antarctic Peninsula where territorial claims overlap, In some other regions many scientific problems are virtually ignored (Lopez 1990, Cross 1991). Although logistic difficulties are a major constraint upon operations, project selection is more frequently a trade off between personal ambition and the necessity for some collectivisation of effort (Elzinga and Bohlin 1989). There is also a certain amount of arrogance within the scientific fraternity, with some researchers claiming an unfettered right of access to Antarctica and no external scrutiny or regulation (Laws 1991). Taken overall, not all scientists are effective advocates for their cause (Cullen 1990).

But whether one regards the motivations and outcomes of some scientific effort as "good" or "bad", the reality is that virtually *all* nations pursue a broad spectrum of scientific research and some part of this activity is expended in Antarctica, a notoriously difficult continent in which to operate (Polar Research Board 1986). If we are to comprehend what is being attempted and achieved in Antarctic science, a series of national research profiles are desirable, of which this brief paper constitutes an introductory segment. Science in Antarctica should not be regarded as *sui generis* however; scientific activities in polar regions reflect not only human curiosity about particular natural phenomena, but a whole series of domestic and international political, economic and social imperatives as well.

In this brief study of Australian Antarctic research it will be argued that one can recognise three distinct eras, each with particular characteristics and orientation, but all reflecting political and scientific perspectives about Antarctica at the relevant time:

(a) idiosyncratic individualism in the "heroic age" of Antarctic exploration 1890-1945;
(b) hydra-headed science programs within the Antarctic Treaty system 1959-1990;
(c) prospective maturity management of the Antarctic environment in the post-CRAMRA era, 1990 onwards.

The latter interpretation is problematical, although current actions suggest acceptance and implementation of the concept.

Australian Science Research in Antarctica: The Initial Exploratory Phase

The Australian Antarctic Territory (AAT) currently constitutes nearly 42 percent of the

119

continent's surface area and reflects both Australian discoveries and an inheritance of prior British Territorial claims in the region (Triggs 1986). Such sovereignty issues are, of course, currently suspended under Antarctic Treaty provisions of 1959.

Prior to 1939 Australian Antarctic science was generally conducted by private expeditions, where research interests largely reflected individual preferences, with some vague notion of future research exploitation if and when appropriate circumstances might arise (Swan 1961). There had been considerable prior exploitation of marine living resources, such as whales, seals and penguins by commercial interests during the nineteenth century, but conservation measures were not seriously contemplated until the 1920's (Cumpston 1968, Mountfield 1974). Noted explorers such as Mawson and Edgeworth David were primarily interested in geological research, but they too advocated consideration of resource exploitation, coupled with the need for some regulatory measures in respect of living species (Spencer 1984). The only Australian expedition prior to 1939 which might be regarded as possessing significant governmental input and structured research priorities was the BANZARE program of 1929-31, which involved joint British, Australian and New Zealand participation. This is not to denigrate the value and diversity of early scientific effort, which was often of good standard, but rather to recognise that science was principally a justification for private expeditions to be in Antarctica and a means of securing limited public donations towards such enterprise.

Consolidating the Foothold: Station Establishment and Antarctic Science 1945-1980

Phillip Law's book Antarctic Odyssey (1983), coupled with accounts by Scholes (1949), Bechervaise (1961), Swan (1961) and others, provide vivid but highly personalised accounts of the establishment of Australia's principal scientific stations on the Antarctic continent (Mawson, Davis, Casey) and on the sub-Antarctic island dependencies (Heard and Macquarie Islands) during the period 1948-1969. Preoccupation with coastal exploration, inland reconnaissance and station building programs did not entirely inhibit effective scientific research, with the gradual development of programs in earth sciences, glaciology, life sciences and upper atmospheric physics.

IGY and advent of the Antarctic Treaty of 1959 induced the Australian government, operating logistically through Antarctic Division and ANARE (Australian National Antarctic Research Expeditions) to enhance policy direction and funding of scientific research. Nonetheless shipping and financial resources remained scarce and much reliance continued to be placed on the role of allied institutions, such as universities and some federal departments, as well as the State government of Tasmania responsible for administering Macquarie Island. In effect two streams of research were developing: the first involving a scientific section within Antarctic Division, the other networking a diverse range of external participants drawn from academic and other organisations.

During the period 1945-1980 the small Antarctic Division suffered a number of changes in ministerial portfolio, ranging from the Department of External Affairs to the Department of Supply and later the Department of Science. An ANARE Planning Committee operated from 1948 until 1966 and although Law has described it as useful for policy orientation and short-term programming purposes, it does not appear to have been popular with Ministers or bureaucrats and was eventually left to "wither on the vine" (Law 1983). An Advisory Committee on Antarctic Programs (ACAP) was established in 1973 and its Green Paper *Towards New Perspectives in Australian Scientific Research in Antarctica* (March 1975) proposed several new initiatives, including a program to rebuild the Antarctic stations, a marine science program and planning studies for an Australian Antarctic ship. Given budget formats of the time, it is now retrospectively difficult to gauge precisely what the total financial outlays were during the 1970's, since elements outside Antarctic Division jurisdiction were involved.

ARPAC, ASAC and the Consolidation of Antarctic Science Programs 1980-1990

An Antarctic Research Policy Advisory Committee (ARPAC) was established in May 1979, to advise on the development of an effective and balanced program of scientific and exploration activity in the Antarctic and sub-Antarctic. Their initial report (November 1979) contained a number of recommendations about priorities in short and longer-term programs, ranging from physical and life sciences to climate, weather and ocean circulation studies, plus "unique" Antarctic science implying issues relevant only to polar regions of the globe. The second report covering the period December 1979 to November 1981 was much more critical of the Australian government, for failure to provide sufficient resources to carry out high priority programs. ARPAC's third report, covering the period December 1981 to November 1983 was even more outspoken, arguing that the decision to carry out a major station rebuilding program on the Antarctic continent, while desirable, did not address the principal strategic objectives of improved science funding, better transport facilities and enhanced logistic support for overland traverses. Air transport to all three continental stations was essential, moreover increased resources should be allotted to marine science projects (Lyons 1991).

In 1984 ARPAC convened a major conference on the future of Australian Antarctic research. Recommendations of conference participants were salutary. The conference agreed that budgetary constraints were severely limiting science programs and that given fiscal limitations some programs might have to be curtailed or abandoned. ARPAC itself was disbanded soon after and speculation continues as to the basic reason. Perhaps the Committee was too outspoken in its criticisms of government funding, but there also appears to have been some concern that ARPAC had strayed from its scientific advisory role and had become too concerned with logistical operations, as well as commenting openly on some aspects of day-to-day management of Antarctic Division.

ASAC (the Antarctic Science Advisory Committee) was ARPAC's successor in 1985. Its terms of reference were to advise the Australian Government through the relevant Minister on:

(a) the broad thrust of Australia's Antarctic program, including scientific, exploration and support activities (including transport);
(b) priority areas for scientific and technological research, having regard to Antarctica's resource potential and the need for sound environmental management; and
(c) measures to ensure an effective Australian participation in international programs involving the Antarctic.

The terms of reference were specifically drawn to exclude matters directly related to the management of Antarctic Division, but do permit ASAC to comment on aspects of transport and logistics which might affect research programs (Lyons 1991).

ASAC's first report, covering the period September 1985 - December 1987 identified seven priority areas, with several recommendations relating to each:

- Unique Antarctic Science
- Earth Sciences
- Weather and Climate
- CCAMLR
- Technology and Support
- Environmental Management
- Social Sciences

Each year's program is submitted to ASAC for approval after project applications have been assessed by specialist subcommittees collectively known as AREG (the Antarctic Research Evaluation Group). Antarctic Division has some opportunity to comment on proposals and operating as ANARE, is responsible for berth allocations, field transport and other logistical support. There is considerable monitoring of outcomes, with earlier emphasis on other aspects, such as field safety, environmental protection, animal ethics, quarantine requirements and the like.

At face value these carefully structured arrangements operate well, but there are some quandaries and operational problems. First, the overall level of ASAC grant funding is exceedingly thin, the real value of project approval lies more in the area of logistic support for field operations. Second, a genuine query exists as to the appropriate balance and direction of scientific research within Antarctic Division relative to work varied out by a variety of external institutions, especially universities. Third, there is the desirability of program coordination and commissioned work, as against the serendipitous nature of project applications varying from year to year. Fourth, there is the measurement problem of identifying actual budget outlays by all participating institutions, as against the science budget component within Antarctic Division itself.

What does the overall pattern look like at the present time? In brief, approximately 130 projects, grouped into 7 priority areas, are funded and/or approved for logistic support each year. The largest number of projects tend to be in life sciences (46) and glaciology/earth sciences (30) with only 6 projects in environmental management receiving approval in 1990-91. Apart from Antarctic Division, some 28 other institutions are directly involved, principally universities.

In terms of budget outlays, Antarctic Division's component is by far the largest, totalling $62.7 million in 1990-91, in a total Antarctic science expenditure of $68.4 million. Within Antarctic Division, direct science program outlays totalled $8.4 million, with expenditure on shipping and general logistic support of $45.4 million. The ASAC grants program is minuscule in comparison ($0.5 million), but this ignores logistic support already included in the Antarctic Division total and other funding sources, such as ARC grants within the universities. All figures quoted must be treated with some caution, as there are many hidden factors and aggregation is not simple.

Such an internal (national) perspective also neglects the evolving context of Antarctic Treaty deliberations and decisions at the international level. In the early 1980's most nations were still interested in the possibilities of resource exploitation in Antarctica, but reluctant to face a scramble for such resources and thus anxious to put some safeguards in place, hence negotiations to develop a minerals regime (CRAMRA) were in train (Peterson 1980, Vicuna 1983, Beeby 1989). By the mid 1980's the activities of NGOs such as ASOC and Greenpeace were forcing a redirection in favour of conservation measures and the notion of a world national park in Antarctica was at least in contemplation (Barnes 1982, Mosley 1986). Cutting across such concepts were the actions of smaller nations, speaking in favour of the "common heritage" concept in the United Nations and challenging the right of the Antarctic Treaty nations to manage the continent (Beck 1986 and 1989, Hamzah 1987). By the late 1980's Australia and France had rejected the draft CRAMRA convention in favour of a comprehensive environmental protection regime for Antarctica, still leaving some related issues such as sovereignty, liability and tourism to be later resolved (Bergin 1990, Davis 1991).

In one sense these dynamics had passed some scientists by, given their preoccupation with particular natural phenomena seemingly abstracted from political or economic considerations. But in another sense a revolution was brewing in project priorities, since adoption of a comprehensive conservation regime would imply a new spectrum of research priorities in fields such as environmental impact assessment, waste management systems, nature conservation in protected areas and possibly the regulation in some degree of scientific activities themselves.

Focussing Research Priorities: The New Agenda

In the current era of global strategic and economic restructuring, environmental security has gained increased salience as a priority on the political agenda. Within the contentious area of regime design for management of the global commons, Antarctic serves as illustration of an area where multilateral agreement has fostered peace and cooperation (Beck 1990, Harris 1990). But if Antarctica also serves as a virtually pristine scientific reference point in an otherwise polluted world, its value to mankind is enhanced and all nations have a legitimate interest in its governance and modes of scientific investigation (Kimball 1990, Herber 1991). Australia cannot stand aloof from such considerations, since it has actively promoted the abandonment of CRAMRA in favour of Antarctica as a "nature reserve - land of science" and must set an example in environmental management of a standard it would wish other nations to emulate (Bush 1990, Suter 1991).

Moving from the current era of scientific freedom in Antarctica towards more comprehensive regulation and minimisation of human impacts may not prove a simple assignment. Apart from fears amongst scientists that emphasis on environmental conservation will mean a redirection of funding away from existing programs, there is considerable resentment of the notion that some scientific activities may become restricted in various ways. For example, there is the quandary as to whether geological investigations might be construed as minerals exploration, an activity subject to a moratorium for fifty years. But if the current Draft Protocol on Environmental Protection is adopted by the Antarctic Treaty Consultative Parties at their meeting in late 1991, implementation will involve a whole spectrum of issues not yet fully addressed:

(a) It is not yet clear whether the Conservation Protocol will involve major multilateral cooperation and implementation, or rely primarily on individual nations for operational measures within their supposed "zones of influence".

(b) Considerable debate is likely to attend any attempt to establish and operate an international Committee for Environmental Protection (CEP) or a Treaty Secretariat. Their relationship with SCAR (the Scientific Committee on Antarctic Research) is also problematical, as is the evolving role of the latter institution.

(c) Highly complex issues of jurisdiction, liability and sanctions within Antarctic operations remain to be resolved.

(d) It will be necessary to further clarify the nexus between UNCLOS, CCAMLR and the Treaty system, especially with regard to circumjacent ocean around Antarctica.

(e) Comprehensive baseline environmental monitoring programs will be required if climate change and human impacts are to be assessed. No nation currently appears to have such systems in place and disputation may arise about standards, procedures and documentation.

(f) It is currently envisaged that three types of environmental impact assessment will be established and implemented. No agreement yet exists as to what "minor" or "substantial" impacts really mean and many other technical issues remain to be resolved.

(g) The expansion of tourism and private expeditions raises significant issues about degrees of regulation and the ability of Treaty nations to deal with "third party" nationals in areas assumed under their jurisdiction.

(h) Two other aspects that remain to be dealt with in more detail are the management and conservation of sub-Antarctic islands (see Paimpont guidelines 1986) and improved protected area systems within Antarctica itself (Kriwoken and Keage 1987).

What actions has the Australian Government taken to cope with these new demands?

(a) Antarctic Division and the Department of Foreign Affairs and Trade are making intensive efforts to ensure the Draft Protocol on Environmental Protection is adopted and are already investigating both domestic and international implications, in terms of progressive implementation.

(b) Antarctic Division has appointed an environmental planning officer and has instigated training and procedures aimed at improved environmental protection, including some initial environmental evaluations of proposed field operations and other activities.

(c) The Environmental Management Subcommittee of AREG has met recently, to identify needs and opportunities in environmental research for the next three to five years. The Social Sciences Subcommittee of AREG has also identified policy issues arising from the Draft Protocol on Environmental Protection which will need to be addressed.

It is clear that new emphases and new priorities arising from environmental conservation measures are becoming institutionalised, but this necessitates re-education of attitudes and values, as much as some reordering of scientific programs.

A more substantial issue remains, as to whether the entire scientific research program is adequately focussed or not. Some commentators have argued that research carried out in Antarctic Division is too costly and dependent upon the interests of individuals rather than national priorities; Antarctic Division refute such claims as anecdotal and without foundation, pointing out that international reputation, the publications record and role in long-term monitoring render the Antarctic Division's scientific effort crucial. A counter-argument exists that some of the research projects funded outside Antarctic Division are too reliant upon the voluntarism of postgraduate students, are of short term duration and limited focus and do not adequately address some important problems currently under-investigated. In order to analyse where Australian Antarctic science is currently headed, ASAC has appointed a subcommittee to investigate the matter and it is likely to report late in 1991 or early in 1992. Without wishing to pre-empt its prospective findings, it appears probable that the subcommittee will recommend retention of the ASAC grants scheme, with some financial supplementation, but also suggest more emphasis on a program approach in key priority areas, with the appointment of identified team-leaders and commissioned studies to fill important gaps in knowledge. If such improvements are adopted, it is unlikely they will be fully implemented and funded prior to the 1992-1993 field season.

Conclusion

During the past decade attempts have been made to improve priority identification and resource allocation within Australian Antarctic science, but doubt remains as to the degree the program is effectively integrated and managed. Critics would claim that emphasis on "relevance" has meant adoption of short term projects at the expense of more basic long-term monitoring, but it is doubtful that scientists are agreed as to what limited number of ongoing studies should be funded. Although reportage on achievements is now improved, no detailed evaluation of the effectiveness or otherwise of projects has been carried out using specified criteria. Queries also remain as to the freedom or otherwise for Antarctic Division scientists to pursue their aims, relative to national priorities as a whole, given that the Division receives the bulk of science funding. Now a new issue has emerged - namely the need for a shift in research priorities towards environmental management. Thus far only limited emphasis has been granted to this field. It is a moot point as to whether the Australian Government will find new financial resources to meet these needs or whether existing monies will be reallocated - an option that many scientists will strongly resist.

If it appears that Australia has only been partially successful in focussing its Antarctic science programs, this reflect upon other nations. Many appear to be well behind Australia, both in the scope and reputation of their science activities, but even further in arrears in terms of environmental management. But perhaps we need detailed science profiles of *all* Antarctic and Southern Ocean nations before we can be confident about such judgements.

References

Bahm A, "Science Is Not Value-Free", *Policy Sciences,* Vol. 1, 1971, pp 391-396.

Barnes J, *Lets Save Antarctica,* Greenhouse Publications, Richmond, Victoria, 1982.

Bechervaise J, *The Far South,* Angus and Robertson, Sydney, 1961.

Beck P, *The International Politics of Antarctica,* Croom Helm, London, 1986.

Beck P, "Antarctica As a Zone of Peace: A Strategic Irrelevance?" in: Herr R, Hall R, Haward M (eds), *Antarctica's Future: Continuity or Change?* Australian Institute of International Affairs, Hobart, 1990, pp 193-224.

Beck P, "Antarctica Enters the 1990's: An Overview", *Applied Geography,* Vol. 10, No. 4, October 1990, pp 247-264.

Beeby C, "The Antarctic Treaty System: Goals, Performance and Impact", Paper presented at Nansen Conference, Oslo, May 1990.

Bergin A, "Australia and the Politics of CRAMRA", Paper presented at National Conference, Australasian Political Science Association, Hobart, September 1990.

Beneviste G, *The Politics of Expertise,* Croom Helm, London, 1973.

Bush W, "The Antarctic Treaty System: Towards A Comprehensive Environmental Regime" in: Herr R, Hall R, Haward M (eds), *Antarctica's Future: Continuity or Change?,* Australian Institute of International Relations, Hobart, 1990, pp 119-180.

Commonwealth of Australia, ACAP Report, *Towards New Perspectives in Australian Scientific Research in Antarctica,* Canberra, March 1975.

Commonwealth of Australia, Department of Primary Industries and Energy, Joint Statement by Minister for Primary Industries and Minister for Resources, *Research Innovation and Competitiveness,* Canberra, May 1989.

Commonwealth of Australia, Joint Statement by the Prime Minister and Minister for Science, *Science and Technology for Australia,* Canberra, May 1989.

Commonwealth of Australia, Australian Science and Technology Council (ASTEC), *Environmental Research in Australia: The Issues,* Canberra, December 1990.

Commonwealth of Australia, Australian Science and Technology Council (ASTEC), Initial Outline for Issues and Options Paper, *Setting Research Directions for Australia's Future,* Canberra, June 1991.

Cross M, "Antarctica: Exploration or Exploitation?", *New Scientist,* 22 June 1991, pp 25-28.

Cullen P, "Values and Science In Environmental Management", Paper presented at Symposium on Water Management in the Alligator Rivers Region, Canberra, April 1990.

Cumpston J, *Macquarie Island,* ANARE Science Reports, Series A(1), Antarctic Division, Melbourne, 1968.

Davis B W, "Science and Politics in Antarctic and Southern Oceans Policy: A Critical Assessment" in: Herr R, Hall R, Haward M (eds), *Antarctica's Future: Continuity or Change?,* Australian Institute of International Affairs, Hobart, 1990, pp 39-46.

Davis B W, "Rhetoric and Reality in Policy Process: Antarctica As A Global Protected Area", Paper presented at National Conference, Australasian Political Studies Association, Griffith University, July 1991.

Elzinga A and Bohlin I, "The Politics of Science in Polar Regions", *Ambio,* Vol. 18, No. 1, 1989, pp 71-76.

Hamzah B (ed), *Antarctica in International Affairs,* Institute of Strategic and International Studies, Malaysia, 1987.

Harris S, "The Influence of the United Nations on the Antarctic System: A Source of Erosion or Cohesion?", Working Paper 10/1990, Department of International Relations, Australian

Nat.University, Canberra, 1990.

Herber B, "The Common Heritage Principle: Antarctica and the Developing Nations", *American Journal of Economics and Sociology* (in press), 1991.

Heap J, "Cooperation in the Antarctic: A Quarter of a Century s Experience", in: Vicuna O F, *Antarctic Resources Policy: Scientific, Legal and Policy Issues,* Cambridge University Press, London, 1983, pp 103-108.

Kimball L, *Southern Exposure: Deciding Antarctica's Future,* World Resources Institute, Washington DC, November 1990.

Kriwoken L and Keage P, "Antarctic Environmental Politics: Protected Areas", Ecopolitics Conference, University of Tasmania, May 1987.

Law P, *Antarctic Odyssey,* Heinemann, Melbourne, 1983.

Laws R, "Unacceptable Threats to Antarctic Science", (Editorial), *New Scientist,* 30 March 1991.

Lopez B, "The Cold Clear View From the South Pole", *Dialogue,* Washington DC, No. 1, 1990, pp 26-32.

Lowe I, "The Politics of Long-Term Issues", Paper presented at National Conference, Australasian Political Studies Association, Griffith University, August 1991.

Lyons D, *Organisation and Funding of the Australian Antarctic Program* (in press), IASOS, University of Tasmania, 1991.

Meynaud J, *Technocracy,* The Free Press, New York, 1964.

Mosley J G, *Antarctica: Our Last Great Wilderness,* Australian Conservation Foundation, Melbourne, 1986.

Mountfield D, *A History of Antarctic Exploration,* Hamlyn, London, 1974.

Peterson M, "Antarctica: The Last Great Land Rush on Earth", *International Organisation,* Vol. 34, No. 3, Summer 1980, pp 377-403.

Polar Research Board (USA), *Antarctic Treaty System: An Assessment,* Proceedings of a Workshop, Beardmore Field Camp, Antarctica, January 1985, National Academy Press, Washington DC, 1986.

Quilty P, "Antarctica As A Continent for Science", in: Herr R, Hall R, Haward M (eds), *Antarctica's Future: Continuity or Change?,* Australian Institute of International Affairs, Hobart, 1990, pp 29-38.

Roots E F, "The Role of Science in the Antarctic Treaty System", in: *Antarctic Treaty System: An Assessment,* National Academy Press, Washington DC, 1986, pp 169-184.

SCAR-IUCN, *Conservation of Sub-Antarctic Islands,* Report of a Workshop, Paimpont, France, September 1986.

Scholes A, *Fourteen Men: Story of the Australian Antarctic Expedition to Heard Island,* Cheshire, Melbourne, 1949.

Slatyer R, "Improving the Dialogue Between Science and Government", Address to Royal Australian Institute of Public Administration, Canberra, 1991.

Spenser C, "The Evolution of Antarctic Interests" in: Harris S (ed), *Australia's Antarctic Policy Options,* Monograph 11, Centre for Resource and Environmental Studies, Australian National University, 1984, pp 113-129.

Suter K, *Antarctica: Private Property or Public Heritage?,* Pluto Press, Sydney, 1991.

Swan T, *Australia In The Antarctic,* Melbourne University Press, Melbourne, 1961.

Triggs G, *International Law and Australian Sovereignty in Antarctica,* Legal Books Pty. Ltd. Sydney, 1986.

Vicuna O F (ed), *Antarctic Resources Policy: Scientific, Legal and Political Issues,* Cambridge University Press, London, 1983

CHAPTER 14

Environmentally Driven or Environmentally Benign Antarctic Research?

*R.B. Heywood**

Introduction

Science is an expression of Humankind's curiosity about its environment. A venture into scientific research is a true venture into the unknown. The "obvious" is rarely so! An initial line of research usually leads into further lines of research, and an understanding of natural phenomena is haltingly and slowly acquired. The "needs" of the science drives the research. The good scientist does not find this a frustrating but rather a very exciting and intellectually stimulating process. He/she readily embraces the firm intellectual self discipline and strict adherence to protocols of measurement and analysis that science demands for the attainment of true knowledge and understanding.

The Antarctic Treaty recognised this essential spirit of science, and sought to promote research in Antarctica by allowing complete freedom of investigation, encouraging international cooperation, and requiring free exchange of scientific plans and data. Research carried out under this regime has clearly revealed the major role that Antarctica has played in global processes such as climate and sea-level regulation since its formation on the breakup of the supercontinent Gondwana about 160 million years ago. Its rocks and ice sheets hold vital information on the past which is relevant to understanding present global phenomena and to predicting the effect of future changes such as global warming. Near-pristine environmental conditions provide a vital laboratory for studies on ozone depletion, UV-B irradiation on biota and for monitoring global pollution. The Southern Ocean has the potential for being a major protein source to an ever-increasing world population. It could be utilising a significant portion of atmospheric carbon dioxide and therefore be a factor in controlling the rate of global warming.

The recent debate on the Antarctic Treaty System (ATS) Convention for the

Regulation of Mineral Resource Activities and the subsequent negotiation of a Protocol on Environmental Protection to the Antarctic Treaty, has been overheated by a campaign of misinformation, deliberately waged by environmentalists and willingly aided by even the popular science media. The comments frequently made implied that Antarctica had been ravaged in the name of Science, and that there should be a New Order which would down-grade the status of science in Antarctica. The Antarctic and Southern Ocean Coalition group of environmentalists have stated that all human activities, including scientific research programmes, should be considered harmful to the Antarctic environment until proven otherwise. A very bureaucratic regulation was proposed that would certainly have the effect of severely restricting the amount of research carried out. It would also require scientists to be involved in extensive monitoring programmes which had the real risk of including irrelevant parameters. Scientists had good cause to develop a fear that a future science programme for Antarctica would be set by lawyers and diplomats under the influence of a group of environmentalists who, although vociferous, have little direct knowledge of either Antarctica or science! Such "environmentally (environmentalist) -driven research" would not be acceptable to scientists. The implementation of such a programme would lead to a considerable loss of first class scientists from Antarctic research, which would have profound implications given the crucial value of the latter in understanding global phenomena.

Fortunately wiser council has prevailed. The Antarctic Treaty System has always considered Antarctica to be a Special Conservation Area and has been particularly conscious of the need to protect the fauna and flora and to minimise the impact of Humankind. Over half of the 175 recommendations of Measures that have been adopted over the past 30 years deal with conservation and environmental protection. The need to rationalise and consolidate these Measures into a more formal system was recognised, during the recent Treaty Meetings in Vina del Mar, Chile and Madrid, Spain. The resultant Protocol on Environmental Protection will be adopted at the Bonn, Germany, meeting in October 1991. The Protocol leaves the regulation of scientific research in Antarctica, and its impact on the environment, to the international scientific community. There is good reason to assume that this duty will be carried out conscientiously.

Leading nations active in Antarctic scientific research are fully cognizant of the need to investigate environmental problems which are of global as well as of regional concern, and that it is crucial that this research is carried out in ways that are environmentally benign. No good scientist willingly works in a dirty and untidy laboratory, be it a room in Cambridge or the West Antarctic Ice Sheet, which is contrary to what some environmental activists would have the General Public believe. Consequently the new Protocol on Environmental Protection to the Antarctic Treaty should, and will, be welcomed as a significant and timely development in line with current knowledge of the impact of Humankind on the Global environment.

Furthermore, it is recognised that the hostile environment of Antarctica is no excuse for poor quality research. The criteria for planning the work and for assessing quality of achievement should be applied as stringently as they are to research carried out under other climes. Indeed the importance of Antarctic research to the understanding of global processes, and the high cost of implementing Antarctic activities, demands that this be so.

In support of this contention I will describe how the Research Programme of the British Antarctic Survey is formulated and implemented, and relate this to the new Protocol, particularly to the requirements of environmental impact assessment.

BAS Science Programme

The British Antarctic Survey (BAS) is responsible for almost all the United Kingdom's scientific activities in the Antarctic, particularly in West Antarctica. BAS is carrying out a very comprehensive programme of first class research of global importance in Antarctica with minimal impact on the environment.

The rapid expansion of British Antarctic research in the 1980s made necessary a re-structuring of the BAS scientific programmes and long-term strategy which was published in 1989 as "Antarctica 2000". Five principle themes provide an overarching framework for 16 science programmes . They have been developed with specific regard to exploiting the scientific uniqueness of the Antarctic, and the relevance of Antarctica's global role. They address major areas of research on the interdependent physical, chemical and biological processes of the Antarctic system - a formulation which parallels that proposed for the ICSU International Geosphere Biosphere Programme (IGBP), and take account of widely perceived priorities for investigation in Antarctica and the values ascribed to the assessment of timely and first class contributions to problems of fundamental concern.

The five themes are:

- Pattern and Change in the Physical Environment of Antarctica
- Geological Evolution of West Antarctica
- Dynamics of Antarctic Terrestrial and Freshwater Systems
- Structure and Dynamics of the Southern Ocean Ecosystem
- Physics of Solar-Terrestrial Phenomena from Antarctica

Pattern and Change in the Physical Environment of Antarctica.

Under this theme, five research programmes combine the expertise and technologies of meteorologists, glaciologists, geologists and geophysicists. Field observations and numerical modelling investigate the influence of Antarctica on global weather systems. The related climatic history is being reconstructed from data on the natural chemical composition and temperature of the atmosphere over the past several hundred thousand years, contained in the chemicals trapped within the ice sheet. The objective is a better understanding of the operation of, and coupling between, components of the ice-ocean-atmosphere systems. Geological studies extend climate information over even longer time scales and address wider aspects of climate change such as the effect on, and evolution of, biotas in high southern latitudes. Contemporary processes such as ozone depletion and the "greenhouse" effect are also studied.

Geological Evolution of West Antarctica

The research under this theme seeks a more precise understanding of the ancient supercontinent Gondwana in which Antarctica was the central component, and of the break-up, and subsequent dispersal and evolution of the fragments. Geological work is complemented by oversnow and offshore geophysical studies. The knowledge gained is making a fundamental contribution to many aspects of geological science. Similarly the research is improving understanding of active subduction zones worldwide.

Dynamics of Antarctic Terrestrial and Freshwater Systems

The relatively simple terrestrial and freshwater ecosystems of Antarctica facilitate research on environment-biota relationships. Recently de-glaciated ground permits study of the processes of primary colonisation. The research is combined in a programme which investigates the survival strategies of organisms facing the hostile Antarctica land environment, and relates them to ecosystem development. Data on the various ecosystems is organised within an integrated resource centre. They form baselines against which future change may be detected, be it through global warming, increased UV-B irradiation or more direct human interference. The database is essential to the development of conservation policy and management schemes for protected areas.

Structure and Dynamics of the Southern Ocean Ecosystem

Field and laboratory research are combined with numerical modelling in an investigation of the Southern Ocean Ecosystem, which encompasses all the main components of the plankton and nekton and their abiotic environment, as a means to evaluate possible impact from natural and humankind-induced changes. The focus is mainly on the groups important as living resources, including the Antarctic krill, squid and fish. Until there is a quantitative understanding of the major energy pathways and the principle interactions between components it will be impossible .

to utilise the living resources wisely and conserve the ecosystem. Aspects of the biogeochemical cycling research are of critical importance in assessing the role of the Southern Ocean in global processes such as carbon cycling and climate behaviour. Birds and seals are top predators in the Southern Ocean foodweb and pioneering research is being carried out on energy and activity budgets of free-ranging pelagic species. Causes of population change are being investigated by measuring fecundity and mortality rates, distinguishing between the effects related to age and experience and to the availability of resources, notably food and space. The three programmes make a major direct input to the Commission on the Conservation of Antarctic Marine Living Resources (CCAMLR), including its Ecosystem Monitoring Programme (CEMP).

Physics of Solar-Terrestrial Phenomena from Antarctica

The charged particles streaming from the Sun, the Solar Wind, interact with and distort the Earth's magnetic field. The Solar Wind is changing constantly in strength and direction, and can create magnetic storms in the outer reaches of the Earth's atmosphere, known as Geospace. Magnetic storms result in light emissions known as the Northern and Southern Lights. They also affect radio communications, perturb the orbits of satellites and damage them by intense radiation. They can also cause major power blackouts, such as the loss of electrical power to most of Quebec in March 1989, by inducing major surges of current in long power lines. The convergence of the Earth's magnetic field in the polar regions provides a "window into space". Antarctica provides an ideal location for the large radar and radio aerial arrays by which these phenomena are studied. By studying magnetic storms it may be possible to predict when they will occur and take steps to minimise their disturbance. Many types of natural radio waves can be studied from the ground in Antarctica. Investigation of their behaviour gives information on the density and structure of ionised gas (plasma) in Geospace. This knowledge is applicable to laboratory research into the generation of electricity by controlled plasma fusion, and into improving radio communications. Humankind is already polluting Geospace in a number of ways yet there is insufficient knowledge to assess the likely consequences. A long-term objective of these programmes is to predict also the effects of anthropogenic disturbances.

There are, in addition, two research programmes that lie outwith the science themes, but serve to promote the high standards of BAS activities.

Humans in Isolated Polar Communities

This research programme is carried out by the medical doctors on BAS Research Stations. In general the research takes advantage of the unique populations of personnel who are similar in age, fitness and diet, and isolated from outside contamination. Investigations include the role of the hormone melatonin in controlling the human body's response to seasonal changes in daylight periods - the "jet lag" effect, and genetic changes in species of micro-organisms using molecular biological techniques. The health care research seeks to improve current practices in preventive medicine, and the content of training courses for expeditionary doctors.

Antarctic Geographic Information and Mapping

This is the most recent development in BAS under which its databases are co-ordinated and improved, in order to exploit better cross-disciplinary links between the research programmes. Links are being established to other national and international databases to support the development of international Antarctic strategies. State of the art techniques, including the powerful combination of satellite imagery and digital terrain models, are being used to integrate a wide range of data in map form.

BAS Science Management Procedure

No research strategy, however wisely conceived, can produce first class science unless it is prudently implemented and, quality control is exercised. The BAS procedure combines external peer review with careful management.

Peer Review

The major elements of BAS research have always been subject to an external review process, the present process involves quinquennial and annual events organised through the Natural Environment Research Council (NERC). At five year intervals detailed proposals must be prepared for each intended research programme for submission by NERC to independent external review by ten or more scientists of world-class stature, drawn internationally. The proposals and referees comments are then considered by NERC Programme Review Groups in discussion with the Director, Deputy Director and scientific staff of BAS. The NERC Programme Review Groups are also independent bodies of eminent scientists. The Chairman of each Review Group submits a report, with the proposals, to the relevant NERC Scientific Committees for comment, and afterwards, with their additional comments, to the NERC Polar Sciences Committee for approval. Finally, the proposals are submitted to the Research Council for ratification. Although prolonged, the process is designed to ensure that the proposed research is timely, relevant and feasible with current techniques and available resources.

The nature and scale of the phenomena and processes being studied under the present BAS strategy determine that it will be long-term. Most programmes will have a duration of at least ten years. However concepts and techniques constantly change making not only quinquennial review but also annual review necessary. The latter utilises the fact that work within each programme is based upon a series of overlapping projects of much shorter duration. These component projects, with their very specific scientific targets and sharply defined annual tasks make assessment easy.

The scientists prepare annual progress reports for their projects which relate the tasks set for the reporting year with the actual achievements, and update the project publications list. Any constraints to progress are identified, along with any extra demands on resources that may have arisen. The report also lists tasks for the coming year. Each project is appraised critically by the Director and Deputy Director in June of each year, assisted by the Project Progress Report and discussion with the Head of each Science Division. Under-achieving projects may be subject to modification, reduction or cancellation at this stage. Following this, the reports are collated according to parent programme and sent to the members of the Programme Review Groups for appraisal. The Groups visit BAS at some time during the period July to September, to discuss on-going research and future plans with individual scientists and BAS Senior Management, before each submits a report to the NERC Science Committees and the Polar Sciences Committee. The Polar Sciences Committee finally reports to the Research Council on the quality of BAS research.

Project Selection

A research topic for a project is defined at least 18 months before implementation, through an iterative process within the parent Science Division between proposer and peers. A Project Proposal is then produced, which attempts to define in considerable detail not only the scientific merits of the proposed work but also the costs in resources - staff, scientific equipment, recurrent costs, logistics, environmental impact and waste generation - over the life time of the project. A feasibility study held within the Science Division judges the new project alongside other proposed new projects and established projects. The project proposal is then submitted for detailed appraisal by the relevant experts in the BAS Logistics Sections and Field Operations Working Group, and by the BAS Health and Safety Officer and the BAS Environmental Officer. Their remit is to judge the feasibility of new projects in terms of the Antarctic environment and present technology, and to appraise the total demand on BAS logistics resources and operational requirements that the proposed combination of new and established projects will make over the following five-year period. Final approval is given by the Director BAS after discussion within the BAS Senior Management team (Director, Deputy-Director, Head of Administration Division, Heads of Science Divisions). The new projects are also discussed with the Programme Review Groups during the process outlined under Peer Review.

Although a recent event, BAS is one of the first national Antarctic operators to appoint an Environmental Officer. He has the responsibility for implementing BAS environmental policy and the provisions of the Protocol on Environmental Protection, as well as the day to day co-ordination and supervision of all aspects of the environmental management of BAS activities in the Antarctic. His job also involves organising the safe and proper disposal of waste. He can seek advice from the BAS Environmental Management and Conservation Committee, which is responsible for developing the BAS environmental management policy.

Scientists from UK Higher Education Institutions and other Government Research Institutes who wish to work in the Antarctic do so mainly through co-operation or collaboration with BAS, and by using BAS logistics facilities. Their research plans are subject to the same review procedures. The same holds true for UK scientists utilising the logistics facilities of other nations because their funding is through the UK Research Councils.

British research in Antarctica is clearly subject to stringent scrutiny at every stage from planning, through implementation, to conclusion. The use of independent review bodies with international membership ensures that the research is timely, relevant and of first class quality, and is either only possible in Antarctica, or best done there. It has the justified reputation of being second to none in quality and productivity, and in being extremely cost effective.

BAS Science and the Protocol on Environmental Protection

The Protocol presents no difficulties for BAS. BAS has always acted according to the belief that Antarctica is a Special Conservation Area, and has always sought to protect the fauna and flora and to minimise the impact of its presence. Members of BAS staff have been prominent on all ATS and SCAR bodies concerned with environmental and ecosystem protection, and have been responsible for some of the initiatives embodied in the Protocol. The use of the multi-tiered peer review system involving independent review bodies and referees drawn internationally should prevent BAS scientists inadvertently breaking the Protocol. Referees and Reviewers will certainly not confuse the scale and methods of sampling involved in geological research with those required by mineral prospecting! The hammer does not equate to the drill; the field trip's box of rock samples does not equate with the 1000s of metres of core attained during mineral exploration drilling at $50m^2$ intervals! They will also not permit analysis of geophysical data to be so prolonged as to arouse suspicions of secrecy associated with commercial interest!

The time required by the Environmental Impact Assessment process should provide no hindrance. The full Comprehensive Environmental Evaluation (CEE) is to take no more than 15 months; BAS works on a minimum lead-in time for research projects of 18 months.

Research requires infrastructure - and this section would not be complete without reference to major capital projects. BAS has just built a crushed rock airstrip at its Research Station on Rothera Point, Adelaide Island. It is sited on land exposed within the last 10 years by a retreating ice piedmont, and devoid of vegetation. There are no seal or bird colonies, although small numbers of animals are occasionally to be found on shore during each summer. In view of the introduction in 1987 of Antarctic Treaty Consultative Meeting Recommendation XIV-2 concerning Man's Impact on the Antarctic Environment, BAS voluntarily produced a draft CEE following special site visits during the 1987-88 season. The draft was circulated for comment to all Antarctic Treaty governments, non-governmental organisations and independent experts. A final CEE was prepared, in the light of comments received and a second site visit, and published in August 1989. Work began in January 1990. BAS invited a representative from Wildlife Link to inspect the site during construction in April 1990. At her request the containment berm round the fuel tanks was extended to ensure full retention in the event of catastrophic spillage; it had been previously designed to meet Canadian Arctic specifications. The same person will carry out a further independent inspection of the airstrip and its operation during the 1991-92 field season. BAS have been monitoring airborne dust, hydrocarbon levels in soil samples and heavy metals in lichens on nearby rock bluffs since before construction began in 1990. The work forms a project within the Ecosystems and Conservation Research Programme. An account of the environmental effects will be published once full aircraft operations have begun.

BAS proposes to rebuild and enlarge its research station on Signy Island, South Orkney Islands. Redevelopment is required to provide modern laboratories and living accommodation, and to cope with increasing demand for places by BAS, university and foreign scientists. To minimise the environmental impact, the new station is to be built on the same site with removal of the old buildings and facilities. An Initial Environmental Evaluation was published in 1990 and circulated for comment to date there have been no adverse comments. Again Wildlife Link have been invited to provide an observer to carry out an independent inspection before, during and after completion of the rebuild programme. BAS research has identified some hydrocarbon contamination from the present Station in marine sediments immediately offshore but this declined to near natural background levels within 500 meters. Marine research on plankton and benthos within the bay on which the Station is sited has failed to detect any adverse effects over a period of 28 years.

There has been a continuous UK presence in the Antarctic Peninsula region since the early 1940s. BAS has inherited a number of disused stations, each abandoned when the apparent scientific need demanded access to areas too distant for sensible dog-sledge travel. Harsh criticism has been levelled at BAS yet it has accepted responsibility for these stations and it has applied for, and received, a special tranche of funds to pay for the necessary demolition and disposal. The work is being scheduled so that it can be done in as short a time as possible but without interfering with the science programme. Present initiatives and achievements should not be negated by past events!

In conclusion BAS has found Environmental Impact Assessment to be a valuable management tool which focuses attention on environmental issues at the appropriate time - during the planning stage. It should not hinder the implementation of first class research or well-designed construction projects.

Concluding Remarks

The effectiveness of any environmental impact assessment will be decided during the preliminary stage when it is dependent on national procedures. In most Antarctic Treaty countries these procedures will reflect the grave concern aroused by Humankind's perceived impact on the environment, both regionally and globally. In the few other Antarctic Treaty countries the procedures will be token. This could quickly prove to be the Achilles' Heel of the Protocol unless wise and patient council prevails both within and outwith the Antarctic Treaty System. The majority of nations will operate with the same earnest intention as the British Antarctic Survey, readily and rapidly modifying techniques and operational procedures to reflect increasing knowledge of the Human impact on environment, fauna and flora. But a few nations will not, either through indifference, or, and most probably, through ignorance or lack of resources. If the Environmental Impact Assessment system is so abused, the question arises as to what should be done? Here lies cause for concern. Attempts to adopt a more swingeing regime, along the lines that all activities are to be considered harmful unless proven otherwise, would at best only serve to reduce the amount of valuable science carried out by all nations because of the considerable increase in bureaucracy that would be involved. It would be naive to assume that there would be an equivalent reduction in logistics activity! At worst, such a regime could lead to recalcitrant nations leaving the Antarctic Treaty System but not the Antarctic! It is folly to assume that agreement would easily be reached, and the means found, to "police" the Antarctic.

I submit that the way forward is through patient counselling within the spirit of Article 6 of the Protocol. It is folly to expect of a Nation a higher standard of science and environmental concern in Antarctica than it is able to exercise at home! Co-operation, the sharing of information, and scientific and technical education are required - not legislation. Contrary to the impression given by the recent, deliberately engineered, misinformation campaign, the Antarctic is not an over-crowded, heavily polluted, small island. It is vast, relatively unimpacted area covering a tenth of the Earth's surface. The two percent of Antarctic that is ice-free represents a land area greater than the United Kingdom. At the height of the Antarctic field season, the total number of scientists and support staff, occupying about 50 research stations, varies between 2000-3000. This very low density of humankind has had a "significant" impact at few of these sites. The major known impacts on Antarctica have been imposed from outside the region - CFCs, heavy metals, radioactive fallout, potential effects of global warming!

This is not a plea for complacency, but for an objective and unemotional assessment of the true cost of Antarctic research; research which must be science-, and not environment-, driven; research which is concerned with global environmental problems and must be environmentally benign; research which is essential for the future well-being of the Earth and Humankind.

CHAPTER 15

Some views on Antarctic Research

Rita R. Colwell

Internationally, Antarctic research falls under the perview of the Scientific Committee for Antarctic Research (SCAR), which belongs to the ICSU-family. In 1991 the General Committee of the International Council of Scientific Unions listed SCAR as one of the bodies to be reviewed. For this purpose a panel of experts selected from the scientific community at large was appointed, with myself as Chairperson.

The panel was asked to review Antarctic Research, in-

Rita R. Colwell is the President of the Maryland Biotechnology Institute and Professor of microbiology and biotechnology at the University of Maryland. Dr. Colwell has conducted studies and published *ca.* 400 scientific papers and 12 books in marine biotechnology, marine and estuarine microbial ecology, survival and pathology of pathogens in the marine environment, deep sea marine microbiology, microbial degradations, and release of genetically engineered microorganisms into the environment. She has served as Vice-President for Academic Affairs at the University of Maryland, Past President for the American Society for Microbiology and on numerous NAs/NRC committees in the US, including present Vice-Chairmanship of the Polar Research Board. Dr. Colwell is Past-President of Sigma Xi, a member of the US National Science Board, and President of the International Union of Microbiological Societies.

cluding the goals of SCAR, the quality of its work, financial support, and success of SCAR in the formation of networks, and activities of networks for interdisciplinary international research. The Report of the review was presented at the 30th meeting of the ICSU General Committee in Jerusalem 5-7 November 1992, at which time its conclusions and recommendations were approved. It was recognized that SCAR itself should conduct a detailed appraisal of its role and structure in the light of the points made in the review report. There was general agreement that SCAR should be encouraged to give first priority to science, without discarding its role in respect of policy making, and to consider incorporating Arctic science. Furthermore it was suggested that SCAR reexamine its organization and consider fund-raising for its activities.

In the following, a presentation is given of some views on Antarctic research developed during the course of the review. These are in large measure based upon the aforementioned report. As will be seen, the general thrust is that the the role of SCAR as a lead agency in international Antarctic science is extremely important.

Given the changes in the Antarctic Treaty System which have taken place in recent years, and the considerable pressures driving research agendas into a direction of relevance, i.e., monitoring activities, a carefully defined mission for SCAR is, therefore, imperative.

From IGY to IGBP

Historically, SCAR was established as a continuation of a special committee on Antarctic research whose task was to oversee, coordinate and stimulate research during the International Geophysical Year.

This Special Committee was formed at the Hague in March 1958, on the authority of the Bureau of ICSU, delegated to that body by the Executive Board at its meeting in Brussels, July 1957; the name was later changed to Scientific Committee on Antarctic Research in 1961. At the time scientists from the twelve nations then active in Antarctica were involved in its activities.

The membership of SCAR comprises the National Committees of national scientific academies or research councils of those nations which are active in Antarctic research, relevant ICSU Scientific Unions and Associate Members which are those national scientific organizations planning to become active in Antarctic research. During the past three decades the membership has more than doubled. At the end of his Report for 1990, SCAR's Executive Secretary (Dr. P.D. Clarkson) listed 24 Full Members, seven ICSU Union Members and four Associate Members. At the 22nd meeting of SCAR, 15-19 June 1992 in San Carlos de Bariloche, Argentia, one further Full and two Associate Members were added. In addition the names of a number of senior scientists, known for their exceptional service to Antarctic research in the past, are associated as Honorary Members in their own capacity as individuals. The current President is Dr. Richard M. Laws of the United Kingdom, and Past President is Dr. Claude Lorius of France.

For all intents and purposes, within SCAR, the "Antarctic" is considered to be that area which is bounded by the Antarctic convergence, although certain sub-Antarctic islands which lie outside the Antarctic convergence may be included in SCAR's area of interest (see Appendix III). In other words, it has not been found necessary, either in the past, or at present, to give a more precise definition of the limits of the oceanic areas in which it is interested.

The main purpose of SCAR was, and still is, to provide a forum for scientists of all countries with research activities in the Antarctic to discuss their field activities and plans and to promote collaboration among them.

From the outset SCAR has operated as a network, meeting once every other year, hosted by participant countries on a rotational basis. SCAR has a small secretariat, located at the Scott Polar Research Institute (SPRI) in Cambridge, UK. SPRI publishes *Polar Record*, a journal on polar research which incorporates activities reports relating to SCAR. A *SCAR Bulletin* appears quarterly within *Polar Record*, and includes documents from SCAR meetings and meetings of the SCAR Executive. Since 1986, *SCAR Report* has been issued as an occasional publication to disseminate more detailed special reports from SCAR and its eight disciplinary working groups, as well as a number of specialists groups created specifically for important missions.

During the latter part of the 1980's, in response to increased relevance and accountability pressures, especially in the domain of environmental concerns, SCAR has restructured a number of working groups and created a number of new specialists groups to deal with matters of research related to new initiatives in this domain. A steering group of the IGBP was also created and it has recently been upgraded in a form of a group of specialists to liaise with international research programs and develop six core programs for SCAR. The working group for logistics has been reorganized under COMNAP. SCALOP, with the organization of managers of Antarctic programs (COMNAP, Council of Managers of National Antarctic Programs, created in 1988) is federated with SCAR. SCALOP replaced the former Working Group on Logistics. Unfortunately, there is a problem determining overlapping responsibilities amongst SCAR, SCALOP and COMNAP. A close working relationship and interactive feedback between SCAR and COMNAP/SCALOP is essential, particularly for supporting large international and interdisciplinary programs. The role of SCAR should be one of scientific policy advice, promoting appropriate research priorities for cooperative adoption within the national programs represented by COMNAP.

Current Strategy

In its recent strategy discussion SCAR's Executive has noted the need for the organization to "improve its influence and visibility in Antarctic affairs", especially in the light of "the rapid tempo of change in Antarctic affairs with the emergence of important issues alongside science, legal and juridical matters, conservation and environmental concerns, commercial interests, and the wider influential political framework evolving through the Antarctic Treaty System." It was also noted that there is "a growing need to reexamine the whole question of scientific data and information exchange in relation to global scientific programmes". Development of an Antarctic data base system is currently being addressed by a combined SCAR/COMNAP group which has replaced the earlier SCAR *ad hoc* committee. This is particularly important in the geosciences where many countries have accumulated acoustic and geologic data in the course of offshore surveys. These data are being collected and stored in a Seismic Data Library System, which was developed by SCAR. Guidelines are being discussed, to standardize, but also because some of the data collected by different countries is sensitive from a commercial point of view.

The *ad hoc* Committee on the Coordination of Antarctic Data covers biology, geosciences (including glaciology), atmospheric sciences, geodesy and geographic information systems, BIOMASS (a multicountry program on marine resource data started in 1976), and logistics. Development of a proper data management structure for proposed large programs would be useful. Retrospective development of a data base for past research, while it may be valuable, is not driven by clearly identified scientific priorities and could be viewed as a service function.

Smaller and newer SCAR nation members have also been asking for better coordination, particularly of scientific research programs, in order to minimize costs.

Task Differentiation

Ad hoc groups of specialists have a dual mandate, both scientific and advisory. The Group of Specialists on Southern Ocean Ecology, established jointly with SCOR (the ICSU Scientific Committee on Oceanic Research), for example, acts as a forum for review and coordination of ongoing and new activities in Southern Ocean ecology and related fields and responds to requests for advice on the possible impacts on marine ecosystems from fishing.

The disciplinary working groups include the atmospheric physics group (renamed to cover the "physics and chemistry of the atmosphere") reflecting a stronger focus on environmental factors, like radioactive elements and pollutants while a new group has been created to specialize in the "upper" end, viz., for solar, terrestrial and astrophysical research. The ionosphere and magnetosphere, are now known collectively as Geospace, reflecting the turn to systemic thinking. A group on solid-earth geophysics is concerned with the structure and dynamic behaviour of the Earth as a system.

The glaciology group, concerned with the physics and chemistry of the ice sheet, seeks to achieve a better understanding of climate change and is encouraging investigations into past levels of "greenhouse" gases from the "archives" of ice sheets by deep core drilling. The variability in the seasonal growth and the extent of the sea ice is also significant since it affects the total albedo (i.e., reflective property) of the continent and its climate, with repercussions on world climate. Studies of glacial movement and the interaction between the ice, ocean and atmosphere increasingly use remote-sensing techniques, particularly those available in orbiting satellites. Here transnational collaboration and standardization of databases is important, not least for those analysts who seek to test global computer simulation models of the atmospheric, oceanic and cryospheric systems.

The working group on geology encourages analysis of how the continental system of Antarctica was formed; what are the crustal forces and processes that shaped its past and will continue to shape its future? By studying rocks and fossils the climatic history may be unravelled and thus the evolution of the animals and plants that lived on and around the continent understood. Many of these problems are beyond the logistic and financial capability of any one nation, so SCAR has set up two groups of specialists to promote and coordinate international research in two specific fields: one on the structure and evolution of the Antarctic lithosphere (the rigid shell of the Earth), and the other on the evolution of Cenozoic paleoenvironments (i.e., from about 65 million years ago) of the high southern latitudes.

The SCAR Executive in 1988 created the Group of Specialists on Environmental Affairs and Conservation, a sign of further accommodation to environmental issues. The importance the logistics function has also been upgraded, with the creation of the Standing Committee on Antarctic Logistics and Operations (SCALOP), which is outside but "federated" with SCAR. Other SCAR Working Groups include, for example, WG on Geodesy and Geographic Information, and WG Human Biology and Medicine.

New Initiatives

SCAR has taken on some new initiatives, such as organizing an Antarctic Science Conference in Bremen in 1991. Unfortunately, the BREMEN conference did not achieve creation of greater visibility for Antarctic research amongst political, public information, and environmental science circles, as had been hoped.
As already noted, many new members have been brought into SCAR. Some of these, especially in recent years, are demanding that efforts be made to assist their participation in Antarctic science more efficiently and in greater coordination. The Third World Nations, of course, are an important component of such coordination.

SCAR scientists are seeking to break the traditional isolation of Antarctic science from international research. The Journal, *Antarctic Science* has been a useful step in this direction, although publications in the wider literature may be helpful in this regard. There have been efforts made to integrate more closely with global programs in other fields, especially the global climate program and to interact more with other ICSU bodies.

The reorganization of working groups and specialist groups in response to environmental concerns and commitments to international efforts such as the global change program has been carried out and more attention has been paid to the type of research needed in the Antarctic. It is not certain to what extent the formation of COMNAP and SCALOP, as independent organizations will, take some of the central coordinating ability and activity from SCAR. If this does take place, it will relieve SCAR of certain burdens in situations where SCAR has been pressed from many sides for advice and demands to act more effectively. It would be useful to define clearly the interrelationships in functional divisions of labor among SCAR, COMNAP and SCALOP in order to define more clearly SCAR's role in this new context.

Redefining SCAR's Lead Role under New Conditions

There is no question, but that SCAR's role needs to be redefined within the changing agendas of Antarctic science.

The greatest concern amongst Antarctic scientists was for maintaining a very high quality of science. A stronger lead role for SCAR in this context would be valuable.

International research stations under SCAR's leadership places demands on SCAR's activities. SCAR could play a more central role in determining criteria for locating new research stations. This has been essentially left to individual nations with, on occasion, political and logistic expediency overriding concerns of scientific merit for projects. SCAR could take an active role if high priority, clearly defined, and relevant scientific research is facilitated by an international facility.

As stated by Elzinga, "Certain questions are raised about the interplay between internal, peer review and quality control criteria and external relevance criteria in Antarctic science. The increasing pressure for relevance coming from environmentalists' concerns is having an effect on Antarctic science and the working conditions of SCAR. SCAR's strategy is to take more account of strategic research, but at the same time, maintaining a solid disciplinary scientific basis. Should SCAR be responsible for international research programs to a greater extent, and should national programs follow these integrated programs more so than now?"

Professor Elzinga further has stated that changing prospectives on Antarctica as a natural resource will affect the perception of this icy continent as an object of research. Thus, the complex interplay of the two dimensions – natural resource and fundamental research – require changing agendas on global policy on the one hand and changing trends in Antarctica research on the other, recognizing that these are a product of both internal and external determinance, i.e., factors internal to scientific interests, as well as factors setting a broader frame of reference.

Thirty years ago, when the Antarctica Treaty was established, basic research stood supreme. Ten years ago, political pressures associated with perceived prospects of oil and mineral potential, together with the actual needs of marine resource management, had a significant bearing on activities of scientists, including the way research agendas were defined and defended. At the present time, pressure from environmental organizations and the introduction of a comprehensive conservationist regime places science under an intense scrutiny that preoccupies scientists. Good intentions, it is feared, might put an end to good science.

Under the Environmental Protocol, adopted 4 October 1991 and now in the process of ratification, there will be an Environmental Protection Committee. In it scientists will have an important role. One of its tasks will be to assess environmental impacts and advise ATS Consultative Parties on management plans for specially regulated areas to be decided upon at Consultative Meetings. SCAR is invited to provide input, but its role appears to be a secondary one, since advice is supplied *via* the Environmental Committee. It is mandatory that SCAR clarify its role more precisely in this new context.

Budget and Organization

In general, SCAR does work of good quality for relatively little money. Currently, SCAR's budget is about $200,000-300,000 US per year, about half of what is reportedly needed to satisfy requests of the various groups within SCAR, calls for workshops on interdisciplinary topics, and other demands. This sum of money, in essence is comparable to what some international organizations in other domains spend for a single meeting, with simultaneous translation and other infrastructural costs. In the case of SCAR, the budget is spread thinly among many highly competing activities. The very small budget is a serious problem, especially when the number of participating countries in Antarctic research is increasing and the number of science advisory tasks and demands for SCAR to carry out peer review, along with new initiatives in the field of science have increased.

In general, SCAR has an organizational structure and a mode of operation suited to conditions thirty years ago when SCAR comprised a group of working Antarctic scientists. It is therefore natural that SCAR should reexamine organizational structure, taking into account the current world situation. The nature of SCAR is one of reacting to need, rather than developing a vision of the future and this has become a problem. Many research scientists are asking SCAR to be more proactive and to develop other institutional arrangements which can better deal with current needs and pressures. SCAR's activity in the science policy advisory arena has been somewhat reduced in the last few years, since a number of other bodies serve this function within the Antarctic Treaty Organization. Consequently, SCAR might consider greater concentration on science and less on science advisory functions.

Increasingly, Antarctic research is becoming an aspect in other international programs or research fronts that are interdisciplinary and task oriented. Clearly there is a globalization at the level of theory, with strong advances in various fields. SCAR has difficulty meeting needs that arise where integration of research results into wider disciplinary and strategic studies must be carried out, especially where initiatives derive mostly from other international agencies and associations. Thus, SCAR finds itself continuously in a reactive mode and it is difficult because of the structure of SCAR, as it is presently constituted, to meet current needs cognitively, institutionally, and at the level of research policy.

Scientific Results and Unique Opportunities

Many of the initiatives undertaken within SCAR have provided very useful results. The BIOMASS program completed its work and organized a major colloquium in Bremerhaven, Germany, September 18-20, 1991. The BIOMASS program dates back to 1976, but traces its origin to the second SCAR Biology Symposium held in 1968 in Cambridge, UK, with the establishment in 1972 of the Subcommittee on Marine Living Resources of the Southern Ocean. The First International Conference on Living Resources of the Southern Ocean was held in Woods Hole, Massachusetts, US, in

1976 where the first draft of the BIOMASS document was completed.

The problem with much of the work that has been done is the proliferation of data and developing efficient data banks in Antarctic research is a problem SCAR recognizes and has tried to address .

The increased number of member nations has resulted in a change in SCAR meetings and organization. The somewhat "clubby" atmosphere of Antarctic research scientists meeting to share information has been replaced by much larger meetings and Working Group membership. Occasionally, discipline Working Group meetings become unwieldy.

During the past few years, increased public awareness and concern about global environmental issues, particularly those in the remaining "wilderness areas" such as Antarctica, have brought more attention to polar regions. The Madrid Protocol within the ATS arises from the new international attitude. SCAR is in a unique position to provide the scientific background required for proper environmental management, but there is also concern that provision of advice to service ATS protocols will dilute SCAR's ability to promote fundamental scientific research. Environmental awareness is increasing at the same time Antarctic scientific activity and tourism (which has potential for explosive, possibly unregulated growth) are putting greater pressures on environment.

Additional pressure on the conduct of relevant Antarctic science, and on SCAR, results from the complexity and interrelationships in the large national systems that are involved in Antarctic research. Thus, strong coordination and effective international cooperation is required for modern, large scale, and sophisticated programs.

From the very brief review of SCAR provided here, one could conclude that the structure of SCAR usefully be examined and a reorganization be considered, with stronger infrastructural support, and a set of clear-cut priority tasks, to obtain a more focused thrust and concentration of effort, including quality control (peer review) in relevant scientific fields. Among the many problems that must be addressed are: the complexity of the organization since many new countries have joined; the role of Third World scientists, both in participating in research and in training programs for their benefit; financial requirements for function and the need to seek new sources of funding; maintaining the initiative in the course of collaboration with strong internatioal research programs based in other international or trans-national organizations; attention to developing a sufficient data base system (e.g., an expanded seismic data library system to embrace all scientific data, the SDLS - Scientific Data Library System); and last but not least, the relationship between SCAR and IASC (International Committee on Arctic Science).

An Antarctic Science Foundation might be formed so that a new institutional arrangement could be devised; also, SCAR could increase its focus on marine as well as terrestrial science. The purpose of an Antarctic Science Foundation might be to provide an institution, in itself international, to coordinate, stimulate and raise the funding needed for research in the Antarctic. The suggestion of an Antarctic Science Foundation arises from the belief that a more proactive lead agency is needed to

stimulate agendas for research based on international cutting edge science, as well as to participate more effectively in other international programs such as global climate and to reduce the time between formulation of ideas and implementation of projects, using the most advanced technologies. However, how such a foundation would be funded would have to be addressed.

There is clearly a strong driving force from national interests in polar research and this could be capitalized upon, giving new institutional arrangements, especially in capturing the interest of European countries.

The notion of an Antarctic Science Foundation is an interesting one, but it should not be construed as an alternative setup. Certainly, there is still a role for SCAR to play, to ensure stronger internationalism in Antarctic research. In any case, the formation of an Antarctic Science Foundation under ICSU and affiliated with SCAR might facilitate meeting some of the needs, as well as facilitate dealing with demands for the institution of international research stations in Antarctica to carry out high priority, clearly defined, and relevant scientific research.

Sunrise in Antarctica. Sketch by S. Duse

Summary and Conclusions

In summary, the goals of SCAR include stimulating, coordinating and overseeing research in Antarctica undertaken under the auspices of a large number of national programs, as well as serving the ATS with advice as requested. The former are achieved through working groups and groups of specialists. The latter is dealt with by groups of specialists and individual scientists. The work of SCAR is certainly not completed and will continue. The latter task will undoubtedly become more difficult and conflicts between the research agenda and the advisory agenda will increase in

the future.

In some cases, research conducted under the auspices of SCAR shows a drift toward monitoring undertaken for non-scientific purposes and also some areas of research appear to be isolated from the international research front. In general, SCAR is currently in a difficult situation because of insufficient funds for the tasks it is asked to undertake. Funds are lacking for important new initiatives and criticism has been raised from SCAR members that too much time and effort is spent on science advisory functions, as for example, with the Antarctic Treaty System (ATS) and more attention should be taken to carrying out excellent science. Finally, it is necessary to find a mechanism for improving access to Antarctica by Third World country scientists.

SCAR's mission and goals might be sharpened so that science coordination, stimulation, and quality enhancing functions are clearly defined and greater attention is placed on science, either down-playing or eliminating science advisory functions.

There is no question but that the funding is much too little for what is needed to operate in accordance with SCAR's goals and the gap is increasing rapidly. The formation of an Antarctic Science Foundation to attract funding and to provide a pool of funds for research in Antarctica, as well as strengthening SCAR's role in Antarctica merits exploration.

Clearly, SCAR's participation in addressing major questions undertaken at the initiative of national research programs, the most obvious example being ozone layer research, requires certain working groups within SCAR, as well as its Executive make every effort to modernize and interact more strongly and effectively within international programs, especially those of global change.

Finally, SCAR has been effective but now may wish to consider current needs for Antarctic research, as well as an organizational structure more in tune with international science and should provide leadership for cutting edge scientific research in areas such as ozone layer research, deep ice core drilling, systems modeling for climatological purposes, paleobotany, Antarctic-related biotechnology, atmospheric chemistry, plate tectonics, etc.

SCAR clearly has a significant mission in Antarctica, especially in coordination of research in Antarctica. Most reassuring is that there are new young scientists participating in SCAR activities. A stronger focus of SCAR on science will, no doubt, serve it well in the future.

APPENDICES

APPENDICES

APPENDIX I

Invited Speakers

Dep. Dir. James Barnes
Friends of the Earth
219 D. ST SE Washington
DC 20009
USA

Nigel Bonner
S.C.A.R.
S.P.R.I. Lensfield Road
University of Cambridge
Cambridge CB2 1ER
United Kingdom

FK Ingemar Bohlin
Inst. f Vetenskapsteori
Göteborgs Universitet
S-412 98 Göteborg
Sweden

Prof. Bruce Davis
Inst. of Antarctic and South-
ern Ocean Studies
University of Tasmania
PO Box 25 2C Hobart
7001 Australia

Prof. Aant Elzinga
Inst. f. Vetenskapsteori
Göteborgs Universitet
S-412 98 Göteborg
Sweden

Dr Ronald B Heywood
British Antarctic Survey
High Cross Madingley Rd
Cambridge CB30ET
United Kingdom

FL Lisbeth Johnsson
Statsvetenskapliga inst.
Göteborgs Universitet
Sprängkullsgatan 19
S-411 23 Göteborg
Sweden

Prof. Anders Karlqvist
Polarforsknings-
sekretariatet
Box 50005
S-104 05 Stockholm
Sweden

Prof. Kent Larsson
Institutionen för geologi
Lunds Universitet
Sölvegatan 13
S-223 62 Lund
Sweden

Dr Riita Mansukoski
Min. of Industry and Trade
P.O. Box 230
SF-00171 Helsinki
Finland

Dr Olav Orheim
Norsk Polarinstitutt
Postboks 158
N-1330 Oslo Lufthavn
Norway

Dep. Dir. Paul-Christian
Rieber
CC Rieber & Co A/S
Postboks 990
N-50001 Bergen
Norway

Dr Finn Sollie
Perspektivgruppen for
Nordområdene spörsmål
Ullernveien 38
N-0280 Oslo 2
Norway

Dr J.H. Stel
Netherlands Marine Research Foundation
Laan van NO Indie 131
2593 BM The Hague
the Netherlands

Prof. Jarl Ove Strömberg
Kristinebergs Marinbiologiska Station
PL 2130
S-450 34 Fiskebäckskil
Sweden

Robin Wendelheim
Forskningsrådsnämnden
Box 6710
S-113 85 Stockholm
Sweden

Invited to speak, but could not come

Prof. Wibjörn Karlén
Inst. f. fysisk geografi
Stockholms Universitet
S-106 91 Stockholm
Sweden

Ms Kirsten Sander
Greenpeace
DK-2100 Köpenhamn
Denmark

Further participants

Pia Eliasson

Christian Swalander*

Astrid Swansson

Rapporteur

APPENDIX II

Program: Changing trends in Antarctic research

Monday, September 30

09.00 - 10.00	Registration at Humanisten, Univ. of Göteborg
10.00 - 10.15	Opening and introduction *Aant Elzinga*
10.15 - 11.00	The role of science in the negotiations of the Antarctic Treaty - a historical review in the light of recent events *Finn Sollie*
Thematic issue	The functional role of science in the ATS 1961-91
11.15 - 12.00	Development of the science/politics interface in the Antarctic Treaty and the role of scientific advice *Nigel Bonner*
12.00 - 12.45	Relevance pressure and the strategic orientation of research *Anders Karlqvist*
12.45 - 13.00	General discussion
Thematic issue	Is science in Antarctica facing the prospect of increasing bureaucratization?
14.30 - 15.15	The place of regulation in relationship to science *Olav Orheim*
15.30 - 16.15	The place of science on an environmentally regulated continent *James Barnes*

16.15 - 17.30   General discussion

Reception   (Ågrenska villan)

Tuesday, October 1

Thematic issue   Orientational shifts in Antarctic research agendas -
                   rhetoric or reality?

09.00 - 09.45   Focusing a national research program - the example
                   of Australia
                   Bruce Davis

09.45 - 10.30   Environmentally driven research - is it different?
                   Barry Heywood

10.45 - 11.30   Geoscience - basic research or potential prospecting?
                   Kent Larsson

11.30 - 12.00   General discussion

13.30 - 15.00   Panel and Plenary session
                   Jarl Ove Strömberg
                   Paul-Christian Rieber
                   Riita Mansukoski
                   Jan H Stel

15.15 - 16.30   Panel and Plenary session
                   Jarl Ove Strömberg
                   Paul-Christian Rieber
                   Riita Mansukoski
                   Jan H Stel

The panel will focus on thematic issues raised in earlier sessions as well as explore
some subsidiary topics:
What is the impact of economic, environmental and other pressures on research
agendas? - a concern for planners and managers
Polar science - what is behind the words?
Changing demands on logistics
Antarctic science from an European perspective
International research cooperation - prospects and constraints

APPENDIX III

Geographical locations of Research Stations in SCAR's area of interest

Argentina
Belgrano II,	77°52'S,	34°37'W
Orcadas,	60°44'S,	44°44'W
Esperanza,	63°24'S,	57°00'W
Marambio,	64°14'S,	56°37'W
San Martin,	68°08'S,	67°06'W
Jubany,	62°14'S,	58°40'W

Australia
Macquarie Island*,	54°30'S,	158°57'E
Mawson,	67°36'S,	62°52'E
Davis,	68°36'S,	77°58'E
Casey,	66°18'S,	110°32'E
Heard Island*,	53°06'S,	73°57'E

Brazil
Com. Ferraz.,	62°05'S,	58°24'W

Chile
Cpt. Arturo Prat,	62°30'S,	59°41'W
Gen B O'Higgins,	63°l9'S,	57°54'W
Ten Rod. Marsh,	62°12'S,	58°55'W

France
Dumont d'Urville,	66°40'S,	140°01'E
Alfred-Faure*,	46°26'S,	51°52'E
Martin-de-Vivies*,	37°50'S,	77°34'E
Port-aux-Francais*,	49°21'S,	70°12'E

Germany
G von Neumayer,	70°37'S,	08°22'W

India
Maitri,	70°37'S,	08°22'E

Japan
Syowa,	69°00'S,	39°35'E
Asuka,	71°32'S,	24°08'E

New Zealand
Scott Base,	77°5 1'S,	166°45'E
Campbell Island*,	52°33'S,	169°09"E

People's Republic of China
Great Wall,	62°13'S,	58°58'W
Zhongshan,	69°22'S,	76°23'E

Poland
Arctowski,	62°09'S,	58°28'W

Republic of Korea
King Sejong,	62°13'S,	58°47'W

Russia
Mirny,	66°33'S,	93°01'E
Novolazarevskaya,	70°46'S,	11°50'E
Molodezhnaya,	67°40'S,	45°51'E
Vostok,	78°28'S,	106°49'E
Bellingshausen,	62°12'S,	58°58'W

South Africa
SANAE,	0°18'S,	02°25'W
Marion Island*,	46°52'S,	37°51'E
Gough Island*,	40°21'S,	09°52'W

United Kingdom
Bird Island*,	54°00'S,	38°03'W
Faraday,	65°15'S,	64°16'W
Halley (V),	75°35'S,	26°15'W
Rothera,	67°34'S,	68°07'W
Signy,	60°43'S,	45°36'W

United States of America
Amundsen-Scott,	90°S	
McMurdo,	77°51'S,	166°40'E
Palmer,	64°46'S	64°03'W

Uruguay
Artigas,	62°l 1'S, 58°51'W	

*Stations north of 60°S

INDEX

Alfred Wegener Institute viii, 95, 97
Amundsen R. 103
ANARE 72, 120, 121, 122
Andersson J.G. x
Andersson K.A. 103
Antarctic Science Foundation 4, 147, 149
Antarctic Science Journal 144
Antarctic Treaty 1, 2, 10, 15, 48,
80, 105, 113, 129, 130, 137, 138,
141, 146, 148, 149
ASOC 14, 37, 66, 107, 117, 123, 130
ATCM 42, 43, 66, 105, 106, 107
ATCP 39, 48, 58
ATCPM 2
atmospheric chemistry 149
atomic bomb 34
Australian Antarctic science 126

Bahia Paraiso 66
BANZARE 120
BAS 77, 79, 80, 131, 133, 134, 135,
136, 137, 138
Beck P. 27, 127
Big Science viii, 24, 45, 58
BIOMASS 25, 143, 146, 147
biotechnology 149
Bodman G. 103
Bremen 39, 40, 144
Bremerhaven 146
British Antarctic Survey 36
British-Norwegian-Swedish expedition 34
Brundtland 8

CCAMLR 9, 18, 24, 35, 38, 51, 58, 107,
111, 113, 115, 117, 122, 124, 133
central institute model 71, 75
Clarkson P.D. 141
climate models 44
climatological models 149
Cold war 40, 104
Colwell ix, 4, 140
Committee for Environmental Protection
109, 124

common heritage principle 123
COMNAP 38, 43, 142, 144
Comprehensive Environ-
mental Evaluation 136
Congress 51, 61, 63
conservation regime 123
CRAMRA 35, 58, 61, 72, 74, 106, 107,
123, 124
criteria 23, 75, 77
cryosphere 42, 143

Daniels 34
data 83, 86, 87
data collection 49, 51, 77, 83, 117, 119,
147
Dome C 67
Dronning Maud Land viii, 31, 81, 84
Dufek Massif 84, 85, 112
Dufek Massive
Dumont d'Urville 67
Duse S. x, 103
Dutch model 48, 92, 94, 98

EAMREA 106
ECOPS 96, 97, 98
EEZ 11, 12, 15
Ekelöf E. 103
Elzinga A. 145
Environmental Affairs and Conserva-
tion (Group of Specialists) 144
Environmental Committee 145
Environmental Impact Assessments 46, 55,
59, 60, 63, 79, 80, 95, 105, 123,
125, 136, 138
environmental monitoring 124
Environmental Officer 135, 136
Environmental Protection 38
Environmental Protocol 1, 4, 8, 13, 15,
17, 18, 22, 55, 59, 63, 64, 82, 99, 107,
108, 109, 111, 112, 113, 115, 117, 124,
125, 130, 136, 138, 145, 147
environmental security 124
epistemic criteria 3

ENVIRONMENT & ASSESSMENT

1. J. Rotmans: *IMAGE*. An Integrated Model to Assess the Greenhouse Effect.
   1990   ISBN 0-7923-0957-X
2. H. Briassoulis and J. van der Straaten (eds.): *Tourism and the Environment*.
   Regional, Economic and Policy Issues. 1992   ISBN 0-7923-1986-9
3. A. Elzinga (ed.): *Changing Trends in Antarctic Research.* 1993
                                                   ISBN 0-7923-2267-3

KLUWER ACADEMIC PUBLISHERS – DORDRECHT / BOSTON / LONDON